Bitter Harvest

Bitter Harvest

A Chef's Perspective
on the Hidden Dangers in the Foods We Eat
and What You Can Do about It

Ann Cooper with Lisa M. Holmes

ROUTLEDGE
New York and London

Published in 2000 by
Routledge
29 West 35th Street
New York, NY 10001

Published in Great Britain by
Routledge
11 New Fetter Lane
London EC4P 4EE

Printed in the United States of America on acid-free paper
Design: Jack Donner

Library of Congress Cataloging-in-Publication Data

Cooper, Ann.
Bitter Harvest : a chef's perspective on the hidden dangers in the foods we eat
and what you can do about it / Ann Cooper with Lisa M. Holmes
p.   cm.
Includes bibliographical references and index.
ISBN 0–415–92227–5 (cloth)
1. Food. 2. Food supply. I. Holmes, Lisa M. II. Title.

TX353.C623 2000
614.3—dc21   99–056123

*To all the farmers who grew the food that fed
the generations before us.*

*To all the cooks who prepared the food that nourished
and nurtured our ancestors.*

*To all farmers present and future whose dedication
to the well-being of the world's people is and will always
be measured by the quality and healthfulness of the food
they produce.*

*To all cooks present and future who dedicate their lives to
providing us with creative and artful sustenance.*

*May you all grow together, combining history and the
wisdom of experience to create and perpetuate a sustainable food
supply for generations to come.*

Contents

Acknowledgments

This book was grown and nurtured with the love and help of so many wonderful people that I have often found myself overwhelmed by all the help, support, time, and energy that they gave.

There are three people, however, who truly made this book. I dare say that without their help the project would not have happened. Gary Holleman, to whom this book is dedicated, was a soul mate and an inspiration. Melissa Rosati yet again believed in me, shared my vision, gave support and direction, and helped to gently but firmly edit and shape my various ideas. Lisa Holmes! Without Lisa this would have been a very different book. She more than anyone else made sense of my thoughts, and English of my tortured verbiage, and brought strength, conviction, and words to my often convoluted brain waves.

Denise LeClaire kept me together. She did research, organization, fact finding, more research, and gave me the time and peace of mind to write. Heather Freedman transcribed the numerous interviews, and Doe Coover, my agent, helped to make sure that this book would be published.

So many cooks, chefs, farmers, activists, and even politicians gave freely of their time, energy, support, and ideas—many have since become dear friends. You will hear their voices and stories throughout the book.

Finally, Randi Ziter and the staff of the Putney Inn. Randi, your love and friendship have kept my heart alive. And to the Putney Inn staff, you have kept me inspired and made me remember that the bottom line is feeding people wonderful food, something you all do so well. Thank you!

I dedicate this book to

Gary Holleman

One day in 1995, a couple of years after our chance meeting at an American Culinary Federation convention, I mentioned to Gary that I wanted to write a book about the history of women chefs. He was very supportive and invited me to meet his editor, Melissa Rosati, who loved my idea and gave me a contract. Gary stayed involved most of the way through the writing of *A Woman's Place Is in the Kitchen* and was always there with advice, encouragement, and cyber hugs. He was an inspiration during all those times when I was sure I would never make it through. He always believed in me and made me believe in myself.

After attending the Chefs Collaborative 2000 conference in Puerto Rico, I joined Gary in his passion for sustainable food and agriculture. We talked about writing a book that would address the issues from a chef's perspective. Together we wrote a proposal and pitched it to Melissa, who got the ball rolling on the project that eventually became this book. Then the unthinkable happened. Gary died unexpectedly. It was a tremendous shock to his friends and family, as well as to his extended family of chefs, culinarians, cyber junkies, and the countless others whose lives he touched.

I believe that the book you are about to read would have made Gary proud. I hope it will be part of his legacy and a conduit for all the goodness and energy he brought into this world and shared with his friends and loved ones.

Introduction

This book owes its beginning to three unique individuals and experiences that influenced me in the early 1990s—Gary Holleman and a computer; Roger Clapp and a lamb; and my niece Abby and a strawberry. The first of these was a chance meeting at an American Culinary Federation (ACF) convention. The computer craze was reaching a fevered pitch, and in an attempt to make sense of the Internet, I attended a seminar where I met Gary Holleman. He single-handedly built my on-ramp to the cyberhighway. Gary had varied and passionate interests in current affairs, and at the top of his list were food and the way we eat.

In fact, Gary was one of the founding members of Chefs Collaborative 2000 (CC2000) and was responsible for helping write the charter by which it is governed. His concern for our food supplies was contagious, and he convinced me to attend a CC2000 convention in Puerto Rico. After sitting rapt through several presentations there, I was hooked. I have been a chef my entire adult life, and I often say that falling into cooking at eighteen saved my life—but cooking for and feeding people has done more than that; it has actually given meaning to my life. Yet, like many young chefs, I did not really have a firm grasp on the origins of our food supplies.

In the late 1970s I attended the Culinary Institute of America, where I was trained by a cadre of primarily European chefs who instilled in us the idea that all great food came from Europe. This sense of "superior" food coming from afar became more firmly ingrained when, after graduation, I took my first job with Holland America Cruise Line. In every port, during the two years I worked on the ship, a celebrated local chef came on board and showcased the local cuisine. In my mind, "exotic" became synonymous with great chefs. I believed, without accounting for seasonality or a sense of place, that all truly inspired chefs served their guests food from different lands.

By the mid-1980s American regional cuisine was making a splash, and although many chefs were still looking toward the exotic, some were exploring regional, seasonal foods. In 1990 I became the executive chef of the Putney Inn in Putney, Vermont, where I immediately fell in love with the local products and decided to create menus around them. I continued, however, to rely heavily on specialty items, often imported and often out of season. Enter Gary, CC2000, Puerto Rico, and a new way of thinking. The conference had a profound effect on me, and for the first time in my culinary career I began to understand the importance of local regional food supplies and the meaning of sustainable food choices.

Shortly after my trip to Puerto Rico, I was invited to attend the inaugural meeting of what would become the Vermont Fresh Network, the brainchild of one of my idols, Roger Clapp, deputy commissioner for agricultural development at the Vermont Department of Agriculture. His goal was to create an organization that would bring chefs and farmers together to promote Vermont's bounty in a project funded by local tax dollars. I was extremely excited to be in attendance at that first meeting. One of the items I was hoping to find a source for was local lamb. I spoke with all of the lamb producers and told them that I was looking for locally produced lamb racks—approximately one hundred per week. To my surprise—in retrospect, my naïveté is almost embarrassing—they all told me that they could not provide the lamb racks. They asked, "What will we do with the rest of the lamb?" "That's your problem," I answered, and they rightfully responded, "No, you're the chef, it's your problem." I was at once dumbstruck and enlightened.

I immediately ordered a whole lamb, and when it came in, I proceeded to help Frank, my butcher at the inn, break it down and figure out how to buy and sell the whole lamb, as opposed to the lamb racks. It was an amazing learning experience, not only for myself but also for the staff, the owner of the Putney Inn, and our guests. Today one of our number one sellers at the inn is our lamb sampler. We buy whole lambs and sell an assortment of cuts on one plate for dinner—a sustainable solution that benefits the restaurant, the farmer, and the consumer.

Not too long after my lamb revelation, my young nieces Abby and Brittany and my nephew Justin came to visit me in Vermont, and I decided to take them strawberry picking at Harlow Farm. When I told them how we were going to spend the day, one of my nieces said, "Aunt Ann, I'm too little to pick the strawberries. How will I reach them?" I was shocked that children, and especially the niece of a chef, did not know how strawberries were grown—their only source of berries had been the grocery store.

It struck me then that as a chef it is my responsibility to teach the next generation about our food supplies, sustainability, and seasonality. My own journey toward sustainability and an understanding of how our food is produced has taken many years and has culminated in this book, which I hope will inspire you to understand and be passionate about your food. I hope that at the very least, after reading this you will pay more attention to how your food tastes, where it comes from, and whether it is produced in a way that is sustainable for generations to come.

Ann Cooper
The Putney Inn
Putney, Vermont

· 1 ·

A Brief History
of Food and Agriculture
in America

sus·tain (se-stân[1]) *verb, transitive*
1. To keep in existence; maintain.
   To supply with necessities or nourishment; provide for.

sus·te·nance (sùs[1]te-nens) *noun*
1. **a.** The act of sustaining. **b.** The condition of being sustained.
2. The supporting of life or health; maintenance: *"to deliver in every morning six beeves, forty sheep, and other victuals for my sustenance"* (Jonathan Swift).
3. Something, especially food, that sustains life or health.
   [Middle English, from Old French, from *sustenir*, to sustain. See sustain.]

The definition of the word *sustain* is concrete. Apply it to something as complex and diverse as a food system, and you will find yourself in the midst of heated international debate. Every person involved with ensuring the sustainability of our food and agricultural production systems seems to have a different idea about what it means. For me, it is simply a *way of growing, harvesting, and consuming food that promotes the perpetual health and well-being of the planet, and its inhabitants, in a cycle that returns to the earth as much as it consumes.*

Throughout history the story of food has been inextricably entwined with that of people. We must eat to survive, and our survival depends on our ability to acquire and produce food. For as many as seven to ten thousand years, life on earth maintained an equilibrium. Food was plentiful, its supply seemingly endless. As humankind evolved and hunting and gathering techniques became more sophisticated, the earth's population began to expand. Over the millennia, population increased, and the need arose to grow more food on less land. Again, humankind adapted, intuitively creating the tools and technologies that helped carry the world's people into the twentieth century. In her book *Food in History*, Reay Tannahill writes,

In the late twentieth century, after more than 50,000 years of fully human intellectual and technological development—when chips of mineral from the earth have been taught to do the work of the human brain, when cruising in space and blowing up the world are among the commonplaces of international discussion—it is still possible for an estimated two million people in Africa to die in a few months from lack of food; possible for the entire economy of Europe to be distorted by the farm price of butter; for Thailand to be impoverished by a feedstuffs war between cassava root and barley; for wheat farmers in Australia to be driven bankrupt by subsidies paid to their competitors on the other side of the world; even for the Great American Hamburger to come under threat because the Russian harvest fails.[1]

This staggering dichotomy between technology and basic human survival is the current story of our relationship to food. International trade, war, poverty, and their complex ties to the world's food supply create a multitude of environmental, ethical, political, and personal dilemmas that must be addressed if we are to successfully sustain healthy and prosperous life on this fragile planet as we head into the new millennium.

THE ROOTS OF FOOD AND AGRICULTURE IN NORTH AMERICA

The history of food, as the shelves of any bookstore will testify, is not as straightforward as it may seem. Food is not just food: it is ritual, tradition, and memory. There is a spiritual connection between humanity and the bounty that springs forth from the earth's fields, farms, streams, rivers, and oceans, and it has existed since the first tree-dwelling creature learned to eat with its hands in a way that satisfied hunger and delighted the senses. Until recently, one did not require the aid of a culinary history to comprehend that bond. Unfortunately we now find ourselves in a place where our connection to the land has become so illusory that the only way to discover where we are today is to look back at our beginnings. It is through this exploration of the past that we will understand how we came to find ourselves standing at a crossroad where the road ahead could lead us, and our children, over a precipice.

A WORLD OF HUNTERS AND GATHERERS AND THE ROOTS OF AGRICULTURE

Thousands of years ago, humans were nomadic, their life cycles influenced by the same climatic changes that dictated annual plant growth and animal migration. Their relationship to the earth was markedly dependent on the capricious ways of Mother Nature. Literally at the mercy of the weather, early humans lived by their wits. What the earth failed to yield, they did without. Like the very animals they hunted, they were biologically programmed to follow the rhythm of the seasons. In fact, prehistoric women gave birth only in the spring because hunting yielded insufficient supplies of nutrients in the winter to allow for procreation.

Women were the primary early gatherers, and it was their observations that led to the crude beginnings of an agricultural system. Using only digging sticks as tools, they learned to sow wild grass seeds in a way that yielded a moderately reliable harvest and henceforth became a predictable food source. Through trial and error and an increased understanding of the cycling of nature's seasons, early people learned to grow food for storage and to preserve meat for later consumption. Nomadic wandering was gradually replaced by the settlement of villages that centered on primitive crop management and production. The rise of agrarianism brought more stable and diverse food supplies, as well as the evolution of woman's biological chemistry: "As chemically pre-set breeding with related social behavior changed, so did female/male relationships resulting in longer term relationships as the norm. For women, these changes meant the birth of children throughout the year, out of sync with the bounties of spring."[2] Year-round breeding put new pressures on communities to produce and provide nourishing food. The survival of the species, in fact, depended upon it. This was the beginning of a cyclical process that not only encouraged the production of more offspring, but necessitated it. Early agriculture was extremely labor-intensive; the more people a village supported, the more food its people needed for survival and the more people it required to produce that food. At best, prehistoric population figures are nothing more than an educated guess; but it is believed that between 10,000 and 3000 B.C. the planet's population soared from 3 million to 100 million, an increase widely attributed to this type of agricultural growth.

Knowing that nature could not be outwitted, prehistoric peoples followed seasonal cues without question and learned to cope with unpredictable changes in weather and unreliable soil quality as well as the inevitable pests and predators. They adapted as necessary, passing agricultural techniques, traditions, and rituals down from one generation to the next through direct hands-on experience. Everyone in a village participated in food production. The hardship of tilling the soil made them thankful for the harvest and spurred them to make efficient use of every part of their crop. By and large, food production and consumption was regional, seasonal, sustainable, and organic. The land was venerated, worshiped, and replenished as much as possible, and by 1500 B.C., all major food plants of the twentieth century, with the exception of sugar beets, were already under cultivation somewhere in the world.

American Indians are presumed to have first cultivated almost 50 percent of the world's plant foods. Out of their belief that their success or failure as farmers and hunters was solely dependent upon their relationship with Mother Earth grew a tradition rich in ceremony in which Native Americans such as the Anasazi, Navajo, Zuni, and Hopi worshiped, among others, the spirits of the harvest, the hunt, and the rain. The Hopi believed corn to be the embodiment of the spirit of the Corn Mother. So treasured was the corn plant that when it began to sprout, Hopi tribespeople would "watch each plant grow and sing prayers of encouragement as if to a precious child."[3] In those early cultures men cleared the land, women farmed, and even the smallest child had a job pushing seeds into the ground or watering seedlings as they pushed through the earth. They acknowledged their own hard work, but still believed that Mother Earth was their ultimate caregiver, and it was she they thanked for the harvest. Southeastern Indians expressed gratitude through age-old rituals such as the green corn ceremony, performed each year just before the harvest: "The women began with a ritualistic cleaning of their homes—burning old clothes . . . and making new ones. The ceremony began with building on the communal altar a new fire. . . . After the ceremony the feasting began."[4]

Hunting was equally important to Native Americans. In many tribes a boy's transition to manhood was dependent upon his hunting skills. Some tribes would not allow young men to marry until they were credited with a major kill. As they worked their way to manhood, they were trained to understand the spiritual importance of hunting. Elk, deer, buffalo, bear, salmon, trout, and other food animals were often given spirit status. Ceremonial tradition dictated that prior to setting out, Native Americans must pray to the animal spirits for good luck in the hunt. Likewise, a successful hunt ended in a ceremony thanking the same spirits for sacrificing their lives for the sustenance of the tribe. These early "Americans" were the caretakers of the land that became known as the New World, and by the time the first colonists arrived, they had bred nearly two hundred varieties of maize.

CONQUERING THE NEW WORLD: COLONIZATION, TRADE, AND WAR

Our contemporary relationship with food begins in Europe during the thirteenth and fourteenth centuries, when population growth, scarcity of food among the poor, excesses among royalty, and stories of untold riches in other parts of the world began an era of unprecedented exploration, war, exploitation, and colonization. Without a doubt, monetary rewards and the entrepreneurial spirit were the driving forces of these explorations, but famine in Europe also played a role. Early population growth was easily matched by increased agricultural productivity, but as villages continued to expand and production levels dropped due to overfarming, landowners and farmers were forced to look for fresh, fertile land as well as alternative food sources.

Spice trade profits were the impetus for many early explorations. Middlemen working established overland trade routes inflated prices by as much as 700 percent and became very wealthy. European sailors responsible for bringing the spices to market, who made much smaller profits, believed that bypassing overland routes by sailing directly to the Spice Islands would increase their earnings. Columbus's search for just such a passage brought him to the New World, and Pope Alexander VI decreed that "all imperial rights in formerly unknown and unmapped territories would be divided by drawing a line north to south through the Atlantic allocating Spain all new lands to the west and Portugal those to the east."[5] This would lead to the establishment of slave, sugar, and eventually rum trade.

Spain cultivated sugar in the Greater Antilles as early as 1506, but a decline in the native population led to a labor shortage, which precipitated the introduction of African slave laborers. As competition for sugar and rum intensified, Caribbean islands were monocropped to meet consumer demand. This practice of using the greater part of the land solely for the purpose of export created colonies that did not have the capacity to feed themselves. Brazil and the West Indies arguably became the "first two societies in the world to be dependent on imports for all food beyond the barest necessities,"[6] a foreshadowing of today's "biocolonialism" that will be discussed in later chapters.

The seventeenth and eighteenth centuries brought many settlers from Europe to North America. Early travelers were still hoping to find a "secret" passage to the Spice Islands; the Spanish and French emissaries returned empty-handed, but the English did have some success. When the Venetian navigator John Cabot, exploring for England, reported on the "great cod-fishing banks off the coast of Newfoundland,"[7] fishing fleets from England, France, Portugal, and Holland came racing for the catch. The English were also the first to establish successful colonies in the New World.

In the first English colonial settlements, in Virginia and Massachusetts, winters were harsh and food was scarce. Early settlers were advised to bring at least a year's worth of food and supplies, and as late as 1650 it was suggested that immigrants bring enough for two or three years.[8] From their homelands colonists carried parsnip, turnip, carrot, cabbage, radish, asparagus, green bean, and spinach seeds as well as fruit trees for planting and grafting. In spite of their preparation, they met an overwhelming lack of success in cultivating their first crops, attributable to the harsher sun and harder rains typical of North America as well as to a general lack of agricultural knowledge; after all, most of these settlers were primarily interested in trade, not farming. Fortunately, Native Americans like Chief Squanto, without whose help the Plymouth Colony might never have succeeded, had developed fairly sophisticated cultivation techniques.

With the most primitive of tools, these Native Americans reaped substantial harvests from small tracts of land, and they were more than willing to share centuries-old techniques—such as fertilizing "each hill with three fish heads pointed inward like the spokes of a wheel" during corn planting season—with the British expatriates.[9] In the beginning, the colonists produced nearly all of their food, clothes, and tools on the farm, purchasing only salt. The first groups of New World settlers did carry some livestock along on their transatlantic journeys, but animal husbandry was not a priority. It was much less labor-intensive to hunt wild game, also using Native American techniques. Later, as the colonists acclimated, their knowledge of the new land brought greater success in cultivating their favorite fruits and vegetables. Many of those plants cross-pollinated with native varieties, resulting in hardier, more flavorful or unusual plants than the settlers could have ever cultivated in their countries of origin. As more immigrants arrived, carrying their indigenous species, North America's horticultural diversity expanded, and it seemed that the land would provide indefinitely for the small but growing American population.

Some estimate that 98 percent of the eastern colonies was originally densely forested; for agricultural purposes, colonists cleared a vast proportion using moderately controlled burning. This technique, also introduced by Native Americans, was widely adopted because it required little effort and produced outstanding results. The carbon produced by burning had short-term agricultural benefits, but its negative impact on the local ecology would not be apparent until the mid-1700s. Forest conservation would not become a matter of urgency until the middle of the nineteenth century. Destruction of ground cover caused soil erosion and irreversible displacement of water tables. Mismanagement of land by early farmers who knew little about sustainable farming practices rendered fertile, nutrient-rich land virtually barren within twenty years. It was said to take seven to fifteen years to rebuild soil in a process that required additional labor and significant financial resources. Coastal farmers were able to use fish as an inexpensive fertilizer, but for farmers who were already shorthanded and financially strapped, rejuvenating land simply was not an option; at that time there was so much available land that it actually cost more to put manure or other fertilizers on the fields than it did to buy a new parcel. Incredibly, it was even more cost-effective for farmers to build new barns than to clean manure from the old ones.

Those who moved westward in search of fresh farmland were likely inspired to make the journey by well-spun tales of vast open prairies, game-filled forests, and rivers abundant with fish. For the pioneers who had the strength of mind and body to make a move west, there seemed once again to be more than enough land for everyone. They could not have known that getting there was only the beginning of the battle. Geography to the west was much more rugged than initially expected—the prairies were by no means accommodating. Changes in weather patterns defied prediction, and winters were brutal. To make matters worse, the soil in many areas was much less fertile than in the East and farmers required a great deal more land to accomplish comparable yields on a western farm. To compound the difficulty, as late as the 1700s farmers had not adopted or created technologies to make their jobs easier. In fact, at that time, farming techniques had not advanced much beyond those that had existed 2,000 years earlier. The equipment found on a typical New England farm during the eighteenth century consisted of a plow, a couple of scythes, sickles, a horse collar, hoes, a saw, two wedges, two axes, a shovel, and a wood splitter. Horses and oxen pulled primitive log sleds (a series of logs or planks fastened together to form a carrying platform) and helped with some of the work, but the majority was done completely by hand. Lasting from before dawn until after dark, the work was cruelly exhausting. One man or even a small farming family experienced great difficulty surviving in the West. Family sizes increased proportionately to farm size. Everyone, even the children, did their share. When a family grew too large, children were "put out" to other families who needed help. In exchange, the children received food, clothing, shelter, and possibly education.

By the mid- to late 1700s, 90 percent of all New World residents were involved in farming. During this time the colonies began to develop as an international economic force. Agricultural developments increased to meet the tide of rising commercial interest and considerably eased the way to colonial economic stability. In 1741 the New England colonies exported apples to the West Indies; seventy fishing vessels were working out of Gloucester, Massachusetts, alone. Unique and imported foods came to be prized by the upper class, and newspapers began to print advertisements for unusual food products such as imported orange juice. By mid-century Thomas Jefferson, a staunch proponent of the American farm and an avid gardener, had begun experimenting with European, particularly Italian, species in his Monticello garden. Among the plants he brought to North America were 10,000 vinifera cuttings, broccoli, sorrel, savoy cabbage, Swedish chives, German kale, sour oranges, black pumpkin, and alpine strawberries.[10] Even today, his copious notes remain of great interest to American gardeners.

In 1773 rebellious colonists protested the British-imposed tea tax by heaving chests filled with the offending leaves into Boston Harbor. In 1780, Scottish inventor James Watt developed a steam-powered flour mill, and in 1784 innovative Shakers began selling seeds in small labeled pouches, making all manner of plants available to anyone willing and able to pay for them. By 1787 the Watt steam-powered flour mill had been one-upped by Oliver Evans's first-ever automated flour mill, which was later credited with making white bread widely available to the American public. Three years later, Congress established a patent office that gave inventors protection and added incentive to create products that could be widely marketed for commercial economic gain. The first U.S. patent was awarded several months later to the Vermonter Samuel Hopkins for his improved method of making "pearl ash" (a leavening for bread) and "pot ash" (for soil enrichment). That the very first U.S. patent benefited both food and agriculture was a foreshadowing of things to come. By the end of the decade, Evans had patented his second automated flour mill, the cotton gin had been invented, the *Farmer's Almanac* had had its beginnings, and the Lancaster Road (linking Lancaster, Pennsylvania, to Philadelphia and the Delaware River) had been completed. At the end of the century the United States had won its independence and was well on its way to an era of industrialization whose primary focus was on the economy of commerce.

INDUSTRIALIZATION

The Industrial Revolution began in Great Britain in the eighteenth century, in large part due to an increasingly wealthy population's demand for higher-quality goods and services. The American invention of the cotton gin and the expansion of the cotton textiles industry led the way for future industrial development that slowly found its way across the Atlantic to the United States. The nineteenth century in America was a time that encouraged business, valued commerce, and for the first time in history lured men off their farms in record numbers as they searched for better-paying factory jobs. Urban centers grew quickly as farmland was cleared to make way for city development, and men abandoned their land to be closer to their jobs.

While consumerism as related to food processing and production was already widespread in the 1700s (markets, bakeries, street vendors, and restaurants existed in virtually every town), the family farm had remained the mainstay of American life. The beginning of the nineteenth century saw the phenomenon of widespread urban growth causing a shift away from the family farm. The once nearly solitary, self-sufficient farm was called upon to support a large number of families who had become part of the new urban working class, far away from their agricultural beginnings. Farms were forced to operate as growing businesses, catering to the needs and whims of their new clientele. At the beginning of the nineteenth century it was clear that the face of agriculture would be forever changed.

Inventors set to work creating time- and labor-saving devices that would ease the farmer's burden. The plow, traditionally a wooden implement, was cast in iron in the early part of the century. The new metal plow, which could withstand greater abuse, was met with resistance by farmers who thought cast iron would have a poisonous effect on their soil, and it was not until John Deere began manufacturing the self-polishing steel plow in 1839 that farmers consented to give it a try. By 1857 John Deere was producing 100,000 plows a year. The mechanical cultivator, invented in 1820, weeded corn more efficiently, and in 1831 the McCormick reaper dramatically cut the number of man-hours required for harvesting. As workers abandoned farms in search of gold in the West, thousands of farmers paid $100 apiece for reapers. Their investments paid off. Technological improvements such as these cut the time it took to produce an acre of wheat from 37 hours prior to 1837 to 11.5 hours in the 1840s, and between 1820 and 1860 production increased between 30 and 60 percent. By the end of the nineteenth century, farmers got another boost—this time in the form of a gasoline-powered tractor.

Farmers were not the only ones to benefit from industrialization. In fact, both the dairy and food-processing industries experienced a boom during this period. The centrifugal cream separator eventually led to large-scale commercial butter production, milling machines encouraged efficiency in larger-scale dairy operations, and individual glass bottles boosted sales of home milk delivery. Meats and vegetables were being vacuum-sealed in jars in the early part of the nineteenth century, and in 1825 the first U.S. patent was issued for tin cans, paving the way for widespread commercial shipping and distribution of processed agricultural products. Sugar refining rose to become New York's second largest industry, producing 1,000 pounds of sugar a day to be distributed to some of America's 9.6 million residents, and the country's first spice-grinding operation was incorporated in Boston, Massachusetts. Two other important innovations of this period were barbed wire and the commercial electric range; one would forever change the face of the western lands, the other, the face of food production.

Agricultural, commercial, and economic advances were occurring at an unprecedented rate. To encourage farming, the government distributed $1,000 worth of free seeds in the late 1830s. By the middle of the next decade, Congress, seeking to reduce the cost of imported goods and at the same time increase exports of U.S. agricultural products, became involved in "free trade" agreements that would continue to influence agriculture for decades to come. One of the first of such agreements was the Walker Tariff Act, which gave U.S. consumers access to inexpensive products manufactured in Great Britain; in turn, the British were guaranteed low food prices.

The government also pushed for the continued settlement of western lands in spite of warnings by forward-thinking individuals like agriculturist Edmund Ruffin, who in 1833 wrote, as recounted by James Trager in *The Food Chronology*, that there would be a " 'growing loss and eventual ruin of [the] country, and the humiliation of its people, if the long-existing system of exhausting culture is not abandoned.' He urged, 'Choose, and choose quickly,' encouraging sounder farming practices like 'contour plowing, crop rotation, the use of furrows for careful drainage, and the use of lime and fertilizers to rejuvenate the soil.' "[11] Regardless of the state of the land, the government made western settlement a priority, and in 1852 the Kansas-Nebraska Act, which opened up land previously reserved for Native American settlement by sacred treaty, was signed. This violation was the beginning of the end of Native American agricultural life. Tribes were resettled on expansive reservations, but as immigrant population centers expanded westward, reservation lands were usurped, forcing Native Americans to achieve more production from less land. As agricultural production became less fruitful, and farming ceased to sustain tribal populations, Native Americans were forced to assimilate.

The availability of land seemed to have little impact on the number of farmers abandoning their professions—in fact, by 1853 fewer than 50 percent of Americans were directly involved with agriculture. Still believing that more inexpensive land was the answer, President Lincoln signed the first Homestead Act in 1862, giving American citizens the right to homestead 160 acres of land for only $10. Between 1860 and 1880, nearly half a million people filed applications. At the same time, the Morrill Act appropriated 11 million acres of land to be given to individual states to sell as they wished and guaranteed the establishment of agricultural colleges in every state.

In 1861 Confederate forces opened fire on Fort Sumter, marking the start of the Civil War. The motives behind this conflict were closely tied to agricultural land use: northern states were heavily populated, their farms were small, and advancements that reduced the need for large agricultural labor forces were widely accepted and heavily utilized there. The southern economy, by contrast, was based on large-scale, labor-intensive agriculture, and its source of labor, the African slaves, was incredibly inexpensive. It is not difficult to understand why the North's push to abolish slavery rankled the southern gentry. They simply did not want to put wage-earning laborers on the payroll. They also wanted to expand the practice of using slave labor into western territories acquired during the Mexican War, to which the North was vehemently opposed. Four years after the first shot was fired, more Americans had been killed than in all future U.S. wars combined. Both the Northern and Southern armies suffered from lack of consistent food supplies, and battles were often won or lost by the "full stomachs," or otherwise, of the soldiers. Disease caused by lack of medical knowledge as well as inadequate food supplies killed more soldiers than wounds did, and the southern agricultural industry did not recover from the devastation wrought upon it until the dawn of the twentieth century.

In the early 1880s a dwindling population of farmers was taking on the challenge of feeding America's populace, and although moderate government assistance and farm technology lightened the load, it was still difficult to get products to market. The roads were bad or nonexistent; the cost of transportation was astronomical (at the turn of the century the cost of shipping one ton of goods thirty miles by road was $9—exactly the same cost as shipping the same weight from Europe to the United States); and spoilage rates were high. In 1824 President Monroe directed the U.S. Army Corps of Engineers to begin building harbors, dams, and roadways, and slow progress was made.

In 1804 the first steam locomotive hauled five wagons containing 10 tons of iron and 70 men along a 9.5-mile length of track; by 1840, the United

States had 3,000 miles of track. In the early 1840s, with the completion of the Erie Railroad line, New Yorkers got their first shipment of milk by rail, and live lobster from the Atlantic Coast made it to Cleveland for a ceremonious boiling. It was then shipped to Chicago, where it was tasted and declared by the *Daily American* "as fresh as could be desired."[12] As demand increased, the railroad continued to grow, and by 1850 there were 9,000 miles of track—a 300 percent increase in just one decade. In 1852 the first through train from the East Coast reached Chicago, and a year later, under the direction of the War Department, the government dedicated $150,000 to the investigation of transcontinental rail routes. Prior to the advent of the railroad, cattle were moved from farm to slaughter entirely by hoof, but in 1860 the first livestock was shipped by rail. Chicago, which had already established itself as a transportation hub, quickly became one of the largest grain- and meatpacking centers in the United States. By 1890 the United States had 125,000 miles of track, more than three times that of Great Britain and Russia combined.

Naturally, it was impossible to move food across the country without some sort of refrigeration to prevent spoilage. As a result, the railroad and refrigeration developed virtually hand in hand. Prior to the dawn of the nineteenth century, refrigeration was limited to natural "cold cellars," which were filled with ice cut from nearby rivers and ponds. In most areas where cold cellars were not available, food was stored outside during the winter. In 1802 the first patented "icebox," literally an insulated box filled with hand-cut chunks of ice, was introduced. Gas refrigeration systems would replace iceboxes thirty years later, and by 1860, mechanized refrigerators were commercially available. Shortly thereafter, as demand for cross-country transportation of food products increased, refrigerated railway cars made their appearance. It was then possible to obtain fresh meat and produce from all corners of the nation. As food became readily available from distant sources within the United States, the importance of regional food supplies was diminished, and consumer demand for nonregional products skyrocketed.

It became clear to government officials that American farmers would require a new level of support if they were to survive into the next century. By this time, 4,000 New England farms had already been abandoned, and greater consumer demand, combined with increased speed in production and processing, was putting a strain on those who remained. In 1862 the U.S. Department of Agriculture (USDA) was created, and it immediately began working to improve the business of agriculture through scientific research and development. According to historian Hiram M. Drache, "The objectives of the new department were to collect and to publish useful information; to introduce improved plants and animals; to answer questions of farmers; to test new implements; to conduct tests on soils, seeds, fertilizers, and other materials used in farming; to establish a professorship of botany and entomology; and to establish an agricultural library and museum."[13] In 1866, after a period of study, the USDA issued its first crop report detailing the types and varieties of crops produced as well as the amounts grown throughout the nation. Along with Weather Bureau bulletins (which began in 1869), crop reports became a mainstay of the American farmer. The USDA appointed research experts, set up experiment stations, and began aggressively importing seeds from around the globe; in 1870 it imported 300 varieties of apples from Russia alone. By the end of the century, the Department of Agriculture had grown to be the most powerful force in American agriculture.

During the 1850s farmers banded together to express concern about freight rates, which had been on the rise, making them unable to profitably participate in the marketplace. An offshoot of their initial gatherings was the Granger movement, which led to the founding of the National Grange of the Patrons of Husbandry in 1867. Local branches of this national organization, known as granges, worked to improve the social, political, and economic status of farmers. These granges pressured the Midwestern states to enact statutes known as the Granger Laws. These statutes set out to protect farmers, many of whom considered freight charges on agricultural commodities to be excessive. Eventually, state regulation by statute became regulation by commission, at which point the Supreme Court stepped in, in 1886, and ruled that regulation of interstate commerce could only be imposed by the Congress.

That decision led to federal regulation of interstate commerce and inspired the Interstate Commerce Act, which resulted in a governmental order requiring the railroads to keep their rates reasonable and fair. As a result of increased transportation, greater control over freight rates, and the "good roads" movement sponsored by the USDA, beginning in 1893, food truly became a national commodity.

As the nineteenth century drew to a close, the food industry was experiencing exponential growth. For the first time in history, people began to feel a real sense of distance from their food sources. Processed food products were taking the market by storm, and in 1870 the first U.S. food trademark (#82) was issued by the U.S. Patent Office, to William Underwood & Co., for its deviled ham product. The U.S. canning industry, which processed 5 million cans of food in 1860, increased production to 30 million cans by 1870. In 1873 Nestle introduced Infant Milk Food to the United States, and a year later margarine made its debut. The first canned baked beans, B&M, were produced in Portland, Maine, in 1875. Heinz ketchup, Log Cabin syrup, Aunt Jemima pancake flour, Tootsie Rolls, and Cracker Jacks all appeared before the turn of the century. Coca-Cola had a presence in every state. Along with food processing came increased misrepresentation of product ingredients. For instance, according to figures in North Dakota alone, residents there actually consumed ten times more "Vermont" maple syrup than the state was capable of producing.[14] A pure food law introduced to Congress in 1889 was virtually laughed off the floor, but more serious attention was paid to a petition to hold pure food hearings five years later.

Even technological advances in the latter part of the century had less to do with agriculture and more to do with food processing. A machine to strip corn from its cobs was invented in 1875, and a pea viner was introduced in 1889. Pasteurization became a requirement for milk in many American communities, and vast improvements were made in the canning industry. By the election of 1896, for the first time in our history, returns showed that economic power had shifted away from agriculture.

THE TWENTIETH CENTURY

The Industrial Revolution led the march of technology and innovation right into the twentieth century. Convenience, efficiency, and profit became the watchwords of every business, and agriculture was no exception. Worldwide, by 1900 there were 1 billion hungry mouths begging to be fed, but in the United States added mobility due to railroad development drew young men away from their farms in droves. Farmwork was "too demanding," and industry jobs, which promised adventure, material wealth, and a more comfortable future, seemed infinitely more exciting. Swept into the swirling vortex of city life, these workers did not look back. Census figures from 1900 show that 48.5 million people lived in rural America, while 30 million called the city home. With fewer Americans involved in farming than ever before, land ownership was consolidated. By 1905, just 100 California farmers owned a total of 17 million acres. Fortunately for them, and others who chose to continue farming, between 1900 and 1914 land values and food prices rose sharply, making the early part of the century one of the most profitable in the history of American agriculture.

The city was as far removed from the farm as Americans had ever been, and this new lifestyle required unique methods of food distribution and processing. It became necessary for food to be packaged, stored, and shipped, factoring in distance and perishability. Processors, many of whom held profit above quality, began to use preservatives, additives, fillers, and even rotten or diseased meat in their packaged goods. It seemed that what the American public did not know would not hurt them. It was not long, however, before food adulteration and safety grew to be of great concern to consumers and elected officials alike. Advocates of the Pure Food Law, which was first put forward in the last decade of the nineteenth century, reported in 1904 that

> more than 90 percent of the local meat markets in the state were using chemical preservatives, and in nearly every butcher shop could be found a bottle of "Freezem," "Preservaline," or "Iceine." . . . In the dried beef, in the smoked

meats, in the canned bacon, in the chipped beef, boracic acid or borates is a common ingredient.... Of cocoas and chocolates examined about 70 percent have been found adulterated.... Ninety percent of the so-called French peas we have taken up ... were found to contain copper salts [for color]. Of all the canned mushrooms, 85 percent were found bleached by sulphites. There was but one brand of catsup which was pure. Many catsups were made from the waste products from canners—pulp, skins, ripe tomatoes, green tomatoes, starch paste, coal-tar colors, chemical preservatives, usually benzoate of soda or salicylic acid.[15]

The following year, Upton Sinclair published *The Jungle,* his best-selling exposé of the U.S. meatpacking industry's intentional public deceptions, in which he describes, among other things, insufficient meat inspection, goat meat represented as lamb, minced tripe passed off as deviled ham, sausage containing poisoned dead rats, and lard that, incredibly, contained human remains. Not unexpectedly, meat sales plummeted, finally forcing Congress to take action. In 1906 they passed the Pure Food and Drug Bill, requiring truth in labeling and prohibiting the sale of diseased or decomposed meat, or "dangerously adulterated foods." That same year Roosevelt released a report that inspired passage of a law for a federally funded meat inspection program.

Transportation systems also benefited from the continued growth of technology in the first part of the twentieth century. Distribution of agricultural and processed products from farm to city was the impetus behind much of the modernization that took place during that time. Improved rail transport initiated unprecedented decreases in freight rates; in 1883 the cost of shipping one ton of product one mile was $1.22, but by 1900 that figure had dropped to $0.75 per ton/mile. Steamboats regularly chugged food and industrial merchandise up and down the mighty Mississippi, and by 1910 the national highway system, construction of which had started in 1893, included 1,000 miles of concrete road. The Panama Canal, an awe-inspiring synthesis of technology and engineering, opened in 1914, shortening the sailing distance between the East and West Coasts by over 5,000 miles. A year later the first tanker trucks began transporting milk across the country; by 1920 there were 307,000 miles of surfaced roadway and over a million commercial vehicles crisscrossing the United States.

On-farm technological advancements improved farming life equally. In 1900 the steam tractor replaced horses for pulling threshing machines in wheat fields, and the following year the steam sheller was reported to reduce the time required to shell a bushel of corn by hand from 100 to 1.5 minutes. The wheat combine was able to complete a 160-minute task in just 4 minutes. It was not long before steam engines were replaced by gasoline-powered ones, and by 1908 there were 200,000 cars on the road, 2 percent of which were owned by farmers. The modified Model T Ford became the earliest pickup truck and one of the most indispensable pieces of farm equipment introduced before or since. By 1920 over 30 percent of farmers had cars. Technological improvements in irrigation helped direct water across arid lands that were deemed untenable in 1806, by Lieutenant Pike (of Pike's Peak), who led an expedition to the Rocky Mountains and compared the plains to the "sandy deserts of Africa."[16] By 1900 the world's irrigated land had risen from 20 million acres in 1880 to 100 million acres, and the first sluice gate on the Colorado River helped carry water from Arizona to California. The National Reclamation Act (Newlands Act), passed by Congress on June 17, 1902, authorized the construction of irrigation dams throughout the West, and millions of sun-parched acres became fertile, irrigated cropland, as the natural rivers and watersheds of Colorado's mountains

---

The Evolution of the American Supermarket

Leather merchants George Huntington Hartford and George P. Gilman could never have imagined what their futures held when they opened the first Great American Tea Company store in 1859 at 31 Vesey Street in New York City. Hartford convinced Gilman (at the time his employer) that there was great profit to be made in the Chinese and Japanese tea trade, so the two pooled their assets and began buying entire shiploads of tea at once. Because they purchased such huge quantities, they were able to negotiate better prices and could in turn offer buyers tea at less than one-third the price of their competitors. The red-and-gold-fronted store became so popular that only six years later the Great American Tea Company had twenty-five locations and had expanded to include groceries as well as tea. When stores were added in the West, the two Georges changed the company's name to the Great Atlantic & Pacific Tea Company, which came to be known simply as the A&P. By 1880 there were almost 100 stores in the A&P chain, copycat chains were popping up around the country, and a revolution in American commerce had begun.

Rural country stores lost their customer base as the population migrated from the country to the city, but also because chain stores like the A&P purchased stock in bulk and were able to offer the same products at much lower prices. By 1910 chain stores had captured 20 percent of the retail grocery market. In 1912 the A&P instigated another major industry change by eliminating perks such as credit and delivery. This made it possible to maintain more stores with fewer employees, and soon A&P had a presence in virtually every neighborhood. Stock was purposely chosen for maximum consumer appeal, and although profit margins were narrow, volume was high, making the chain very lucrative. By 1929 the A&P numbers increased to include 15,000 stores, which metamorphosed into larger stores selling meats, baked goods, fresh produce, and groceries. Prior to the introduction of these convenience stores, people had to visit the local butcher, baker, green market, and grocer to do the weekly shopping. Americans had been introduced to one-stop shopping, and they liked it.

The Evolution of the American Supermarket (continued)

By the 1930s 11 percent of all grocery store sales came from the 18,000-store A&P chain, but even as the chain grew, it was closing the smaller "local" stores and opening larger and larger "supermarkets." The first cellophane-wrapped meat appeared in 1940, directing business away from the local butcher's shop, and nearly 15,000 grocery stores were stocking frozen foods, compared to only 500 a decade earlier. Supermarkets generated just under 70 percent of all food sales in 1959, even though they represented only 11 percent of all food stores. By then the typical store carried over 4,000 items—a 300 percent increase in just twenty years. In 1963 the A&P built a 1.5-million-square-foot food processing facility, the largest in the world. They failed to anticipate the mass migration to the suburbs, however, and by 1975 they were forced to close over 1,000 stores.[1]

Consumer pressure continued to make one-stop shopping a must, especially in the suburbs, and stores grew ever larger as the market for convenience items grew exponentially. In fact, since the 1970s new product growth for supermarkets has been more than 11 percent per year, but with a failure rate between 80 and 99 percent. Consolidation of business became the trend in the late 1980s, and by 1992 only four companies accounted for 65 percent of all grocery food sales. The food at these "superstores," and "hypermarkets" is processed and packaged especially to meet consumer needs. In the 1990s the average supermarket carried 30,000 items—a mere fraction of the 1 million products available on the market. Market research suggests that in the future 90 percent of all food purchased will be on the consumer's table within ten minutes of the start of meal preparation—barely enough time to set the table and open the refrigerator door. There is already little reminiscence within supermarket aisles about the wholesomeness of natural, unprocessed foods. Soon everything will be precut, preseasoned, and precooked. If they could just conquer the logistics of pre-eaten food, we would not have to go to the supermarket at all.

1. James Trager, *The Food Chronology* (New York: Henry Holt, 1995), 359–61.

were coaxed to change course. The Newlands Act also provided that individually owned farms of up to 160 acres would have free access to federal water. In 1909 the Rivers and Harbors Bill gave the U.S. Army Corps of Engineers clearance to build locks and dams on American waterways, and in the same year the Laguna Dam on the Colorado River was completed, near Yuma, Arizona. With 3 million acres under irrigation, Colorado became the country's most irrigated state. Within two years, the Shoshone Dam was completed, in Wyoming, and the Roosevelt controlled the flow of the Salt River in the Arizona Territory.

Around the time that irrigation was making farming possible in the country's least hospitable regions, agriculturists began developing an interest in soil health. The use of fertilizers such as manure, fish by-products, and guano to increase production was already quite common, but chemical fertilizers had yet to make an appearance. The U.S. Bureau of Soils undertook its first National Soil Survey in 1909, the results of which led to an announcement that "soil is indestructible." By the end of the year American Cynamid Corporation, founded in 1907, was producing 5,000 tons of nitrogen fertilizer annually. Soil may have been "indestructible," but it was not beyond improvement. During the next decade and a half soil study became an international effort, and in 1924 plans were made for the creation of an International Society of Soil Science. Three years passed before the First International Congress of Soil Science was held, in Washington, D.C.; its first elected president was an American.

WORLD WAR I

Just prior to the onset of World War I there was enormous growth in the processed foods industry. In 1911 the introduction of Crisco, a vegetable shortening, revolutionized Orthodox Jewish cookery because it could be used at any time without fear of violating kosher dairy laws. It would later gain widespread recognition when lard shortages during the war forced cooks to seek an alternative. Mazola Oil, the first corn oil produced for home consumption, made its debut the same year. In 1912 alone, Sun Maid Raisin and Diamond Walnut companies were established, and Ocean Spray Cranberry Sauce, Hellman's Mayonnaise, Prince Macaroni, Oreo Biscuits, Lorna Doone cookies, Quaker's Puffed Rice and Puffed Wheat, Peppermint Life Savers, and the Whitman's Sampler all found their way onto market shelves. Along with these new items came a dramatic increase in food prices. In 1914 Congress passed the Federal Trade Commission Act, which prohibited unfair competition and false advertising. Less than a month later the Clayton Anti-Trust Act, an amendment to the Sherman Antitrust Act of 1890, made its way through Congress. Among other things, the new law prohibited exclusive sales contracts and unfair price-cutting to discourage competition in the marketplace.

The same year, World War I began suddenly, giving an unexpected boost to progress in farming. Beginning with wheat, food production became of paramount importance. Less than fifty years after the potato famine (caused by potato blight due to monocropping) that took the lives of over 1.5 million Irish, Europe was in dire need of supplies. U.S. farmers answered the call by taking 40 million more acres under the plow and initiating a 26 percent increase in production per acre. Even a Hessian fly outbreak that caused $100 million worth of damage to the wheat crop in 1915 did not stop farmers from having a record-breaking year. Prices skyrocketed. In fact, prices in the beef and dairy industries experienced a sixfold increase. The next year, as part of the war effort, the Federal Farm Loan Act was passed to provide credit to farmers for the purchase of machinery and land. Additionally, a Stockraising Homestead Act again doubled land allotments as outlined in the Homestead Act of 1909, increasing them this time to 640 acres.

As the war raged on, food prices continued to rise, and in 1917 poor Americans living in New York, Boston, and Philadelphia were finding it difficult to feed their families. They were told to try substituting milk for meat, and rice for potatoes. At the time, processors were replacing the butterfat in milk with fillers such as vegetable oil, and aside from having been given unsound nutritional advice, less fortunate Americans were angered by the idea that the "land of plenty" had nothing to give them. Riots broke out in all three cities. Food shops were ransacked and peddlers' pushcarts overturned and burned. Fearing havoc wreaked by rioters, independent shop owners also struggled to stay in business as larger stores and wholesale operations were able to offer the same products at lower prices. That same year, as protest against price-cutting, 6,150 kosher poultry shops and slaughterhouses closed their doors just before Passover.

At about the same time, President Woodrow Wilson issued an embargo proclamation, which turned control of exports of food, fuel, iron, steel, and war material over to the government. In a battle cry to all Americans, but particularly farmers, Wilson proclaimed that "Food Can Win the War." In the final year of fighting, the USDA began rationing sugar and announced preparations to declare the booming soft drink industry "non-essential." It was a threat that never materialized, and sugar prices soared. As World War I drew to a close, the United States emerged as a major contributor to the global food supply. In fact, throughout the war effort, it had become the leading exporter *and* importer of agricultural products. With only 4 percent of the world's farmers, the United States had found a way to produce 70 percent of the world's corn, 60 percent of its cotton, and 20 percent of its wheat.

Geographical isolation allowed progress in the U.S. industrial sector to continue virtually unabated throughout the war. Convenience foods and processed products again experienced rapid growth, and before the war ended Americans had been introduced to "Mr. Peanut," Nathan's Famous Frankfurters, Orange Crush soda, StarKist Seafood, Sunsweet Growers products, and Clark Bars. Unfortunately, the end of the war also marked the end of America's economic boom, and farm prices began to fall with amazing rapidity. Within four years they dropped by nearly 72 percent, and farmers again abandoned their land. With over a million fewer farmers, in 1922, for the first time in U.S. history, urban residents outnumbered rural ones. Controlling costs and increasing yields to meet consumer demand became the food industry's primary concern.

In spite of falling prices, some states continued to experience agricultural growth as a direct result of technology. Water reallocations through irrigation made Arizona and parts of southern California important suppliers of fruits and vegetables. (Farmers whose water supplies were drying up as a result of irrigation, on the other hand, were inconsolable. Owens Valley farmers blew up the Los Angeles water gates seventeen times in 1924 to prevent the reallocation of "their" water, which would eventually result in the ruination of their valley.) The first row-crop tractors were improving efficiency, and in 1923 the Erie Canal opened. The time required to move freight from the Midwest to the Atlantic dropped from twenty or thirty days to a mere eight to ten, and freight rates took an instantaneous nose-dive, dropping like a rock from $100 per ton to only $5 per ton. Access to the Atlantic port of New York City through the Erie Canal made commercial success stories out of Buffalo, Rochester, Cleveland, Columbus, Detroit, Chicago, and Syracuse.

In 1920 the agricultural sector believed that increased political power and new governmental regulations were the solutions to its problems. The American Farm Bureau Federation was founded that year in an effort to effectively lobby Congress for agricultural rights and benefits. In 1922 the government responded with the Fordney-McCumber Tariff Act, which gave presidents the power to increase duties on industrial goods to protect sales of farm products. The same year brought the Grain Futures Act, which imposed limits on the types of speculation thought responsible for the 1920 grain price collapse. In 1923 Congress passed the Agricultural Credits Act, founding Federal Cash Credit banks whose purpose it was to provide moderate-term personal and collateralized loans for farmers. In the years to follow, the government continued to pass legislation slated to support the farming industry. At the same time corporate agriculturally based industries began a massive move toward consolidation. The last half of the decade saw three of the world's food giants come into being: Archer Daniels Midland (ADM), General Mills, and General Foods. All three would go on to dominate the food industry through the end of the century.

Even as consolidation was bringing profits to corporate giants at the start of the 1930s, much of the farming community was near bankruptcy, and 4 million workers were unemployed. Breadlines were common in major cities like New York and Philadelphia, and smaller towns in Arkansas and Oklahoma were besieged by hungry rioters shouting, "We want food! We will not let our children starve!"[17] National income decreased from $81 billion to $68 billion, and the government was compelled to step in. In 1930 the USDA established a Grain Stabilization Corporation and appropriated $500 million to buy up surpluses responsible for low farm prices, but it was not nearly enough. At that time, only 25 percent of Americans lived on farms, and the average annual farm income was $400 per family.[18] In 1932 the average weekly relief check in New York was $6, the average weekly wage $17 (down from $28 in 1929), and more Americans were hungry or malnourished than ever before in the nation's history. As nearly 1 million Americans fled the cities and returned to the land, farm prices continued to fall.

The plight of the farming industry had such an enormous impact on the government that in 1932 the USDA spent almost 70 percent of its budget on road construction in an effort to support the farm-to-market connection. At the beginning of the decade the country had almost 700,000 miles of paved road; by 1940 that number had increased to over 1 million miles. In 1933 Congress passed the Emergency Farm Bill to restore farm income by reinstating 1909–1914 average crop prices. Also that year, after realizing that farmers were not going to voluntarily remove land from production, Congress established the Commodity Credit Corporation within the USDA to buy up farming surpluses and pay farmers for taking their land out of production. The Farm Mortgage Financing Act of 1934 was initiated to help keep farmers out of bankruptcy, and the next year brought the establishment of the Rural Electrification Administration, which began a process that would electrify most U.S. farms by 1950. Government continued to finance irrigation projects, and by the end of the decade Hoover, Parker, and Shasta Dams provided water for the population of Los Angeles and irrigated over 250,000 acres of new agricultural land. The Parker Dam alone tapped 1 billion gallons per day of water from the Colorado River and moved it across 330 miles of desert and mountains to the people of Los Angeles.

The government was working at a furious pace to keep U.S. agriculture economically viable. When a dust storm in 1932 swept South Dakota soil as far east as Albany and just two years later 300 million tons of Kansas, Texas, Colorado, and Oklahoma topsoil blew into the Atlantic, they redoubled their efforts. At least 50 million acres lost all topsoil, another 50 million were very nearly destroyed, and 200 million were seriously damaged. Dust storms that blew up after an extended period of drought in the Great Plains states (1934–1936) were so dramatic that farmers were literally forced from their land, and the government stepped in to help farmers stay the erosion. On June 28, 1934, Congress passed the Taylor Grazing Act to prevent over-grazing and soil deterioration, and in 1935 President Roosevelt signed the Soil Conservation Act into law and appointed Hugh Bennett to head the Soil Conservation Service. Much of the topsoil loss was blamed on aggressive farming practices during World War I, when virgin land was plowed and planted with wheat to take advantage of high grain prices caused by wartime food shortages. Bennett estimated that soil erosion was costing $400 million in decreased productivity in the agriculture sector. By the end of the decade Congress had passed the Soil Conservation and Domestic Allotment Act, which paid farmers to plant cover crops of alfalfa and clover in place of traditional, soil-depleting crops such as corn, cotton, and wheat. In the end, farmers were paid not to farm 160 million acres, or one-third of the land tilled in 1930.[19] In 1938, Secretary of Agriculture Henry Agard Wallace (elected vice president of the United States in 1940) attributed U.S. agricultural misfortune to

(1) the wartime expansion of 40 million acres; (2) the change of the United States from a debtor to a creditor nation, which made it more difficult to export; (3) the displacement of horses by tractors, which eroded the value of horses and released more land for market production; (4) the fact that Europeans worked harder to produce their own food; (5) new competition from Argentina and Australia; (6) higher U.S. tariffs, which caused other nations to shut out our products; and (7) the growth of big businesses that could set the prices of what they bought or sold. Only the expansion in acreage and the switch to tractors were directly influenced by farmers; the other five causes were beyond their control.[20]

In the middle of a crisis they did not create, farmers readied themselves for another decade of heartache and difficulty as the 1930s drew to a close.

WORLD WAR II THROUGH THE END OF THE CENTURY

Farmers were granted a reprieve with the onset of World War II, in 1939. Two years later, when the war exploded into a truly global conflict, virtually all able-bodied young men were drafted, leaving a severe shortage of farm laborers on U.S. soil. The war demanded an enormous increase in agricultural production, and just as they did during World War I, farmers rose to the occasion. This time, however, very little new land was brought under cultivation. The weather was optimal, and farming was made significantly less labor-intensive by increased use of power machinery, commercial fertilizer, insecticides, and herbicides. Once again, technology came to the rescue. Wartime production of staple crops, as well as livestock, hit record highs.

The war machine created a need for better engines, vehicles, food processing techniques, roads, and communication. At the conclusion of the war, a great deal of technology designed for combat was transformed for peacetime use in the private agricultural sector. Tank treads became tractor treads; jeeps and military transport vehicles were transformed into the pickup trucks and eighteen-wheelers we see on the road today; submarine sonar became fish "finders"; and battleships metamorphosed into floating fish-processing plants. This type of naval technology brought international markets and businesses that much closer to U.S. shores. Finally, high-energy army rations, which required a long shelf life, spurred the rapid growth of the processing industry, and chemicals originally manufactured for bomb production became fertilizers.

DDT, originally used to protect troops from malaria during World War II, played an important role in winning the war and went on to have a major impact on agriculture. The fertilizer industry grew from over 800,000 pounds during this era to over 5 million pounds within thirty years. Nitro-

Cod

A Story of Unfathomable Human Carelessness

Reddish brown in color, this white-fleshed bottom feeder seems, but for the single barbel upon its chin, a rather unremarkable creature. Hardly. For centuries cod has been nothing short of revered by fishermen, traders, explorers, politicians, and consumers alike. In fact, cod have survived for more than 10 million years, not to mention numerous ice ages. Men have lived and died, economies succeeded and failed, by the fins of this humble fish. Basque cultural history includes a lively legend of a talking cod, and for centuries a wooden effigy of the fish has occupied a place of honor on the wall of the House of Representatives in Boston, Massachusetts. Tall tales and incredible true stories have given this fish a place in history like no other. Its reputation, it seems, is well deserved.

We know from archaeological evidence that Vikings traveling a route from Norway to Canada ate cod. In fact, they probably survived on it. Ocean travel was dangerous, but the Vikings were known for their ability to withstand adversity. They might have been forced nevertheless to choose between starvation and death at the hands of enemies while foraging in unknown territory were it not for the ease with which cod could be preserved. Nailing their catch to planks and allowing it to dry in the cold northern air gave them a necessary and reliable source of protein. By the ninth century A.D. the Vikings "had already established plants for processing dried cod in Iceland and Norway and were trading the surplus in northern Europe."[1] Basque explorers later discovered that the use of salt in the drying process extended shelf life and made more extensive sea travel possible. It was they who brought cod to international markets in the eleventh century.

By the fifteenth century the importance of cod as food and as a source of commercial income was so well documented that when John Cabot discovered "New Found Land," he reported on the abundance of cod "as evidence of the wealth of this new land." Two centuries later the city of Boston was literally built upon the backbone of the cod-fishing industry, and it is believed that the success of the Massachusetts Bay Colony can be directly attributed to the cod catch. In 1640 alone the colony "brought 300,000 cod to the world market."[2] When fishermen suggested hanging a wooden carving of the cod in a place of honor, not one man presented a challenge. It later became known as "The Sacred Cod." Those Bostonians whose lineage could be traced back to the fisheries established in the seventeenth century were proudly dubbed the "codfish aristocracy."

In the early 1800s, despite vigorous fishing of the "banks" (the Atlantic coastline running from Newfoundland to southern New England), cod remained plentiful. In fact, it was still quite common to catch fish weighing in excess of 150 pounds. Even in the latter part of the century the cod population was considered impervious to attacks by man and nature. In *Le Grande Dictionaire de Cuisine* (1873) Alexandre Dumas remarked, "It has been calculated that if no accident prevented the hatching of the eggs and each egg reached maturity, it would take only three years to fill the sea so that you could walk across the Atlantic dryshod on the backs

Cod (continued)

of cod." Around mid-century, however, fishing methods began to change, and the cod population was quickly diminished. Hand fishing yielded to a 1,300-foot net, known as a "purse seine," and fish were literally scooped from the ocean. In the 1920s trawling became the method of choice, and onboard ice-making machines that reduced the risks of spoilage associated with longer trips enabled fishermen to extend their time at sea and increase their catches.

When factory ships equipped with sonar, processing and freezing facilities, and a holding capacity of 4,000 pounds sailed onto the scene in the 1950s, the technology finally rose above the ability of the resource to replenish itself. This seems impossible, given the fact that one codfish is capable of laying over 3 million eggs at a time. In spite of repeated warnings detailing the fragility of the ecosystem, people were so unwilling to believe in the mortality of this godlike creature that they just kept fishing. To protect coastal fisheries and local economies, governments began imposing coastal territorial limits. The first, in the 1950s, extended three or four miles from shore. Following the imposition of the three-mile limit, Britain and Iceland became locked in a territorial struggle that ended in the 1970s after an Icelandic vessel blew a hole in the hull of a British trawler. Eventually, both countries finally agreed to a fifty-mile limit. At a 1973 United Nations meeting, thirty-four countries, including Iceland and Norway, voted in favor of a 200-mile limit. Unfortunately, the cod population in Icelandic waters was already in decline.

Americans and Canadians, who came to blows over who would control the Georges Bank fishery, were each granted a portion of the prized territory and wholeheartedly embraced the 200-mile limit laws. They were not concerned about preservation as much as they were interested in having all the fish to themselves. It was a free-for-all in places like Georges Bank, where dragnetting in the 1970s and 1980s flattened the ocean bottom and rendered the healthiest ocean breeding ground virtually lifeless by the century's final decade. In *Cod: A Biography of the Fish That Changed the World*, Mark Kurlansky wrote,

> In January 1994, a new minister, Brian Tobin, announced an extension of the moratorium [on fishing the northern cod stock (first issued in 1992)]. All the Atlantic cod fisheries in Canada were to be closed except for one in Southwestern Nova Scotia, ... Canadian cod was not yet biologically extinct, but it was commercially extinct—so rare that it could no longer be considered commercially viable. Just 3 years short of the 500-year anniversary of the reports by Cabot's men scooping up cod in baskets, it was over. Fishermen had caught them all.

It is almost unbearable to consider the magnitude of this accomplishment.[3]

1. Mark Kurlansky, *Cod: A Biography of the Fish That Changed the World* (New York: Walker and Company, 1997), 21.
2. Ibid., 28, 70.
3. Ibid., 186.

gen plants, originally constructed to supply the military during World War II, became postwar fertilizer production units. The USDA estimated that 55 percent of increased production per acre between 1940 and 1955 was due to fertilizer usage.

This was the beginning of the "green revolution," which involved breeding crops to produce high yields and developing intensive cultivation methods to be used in concert with "miracle" chemicals. These new techniques saved farmers both time and money, and it was believed that the green revolution would help create the technology that would make feeding the world's ever-increasing population possible. The science of genetics really began to have an impact on plants and animals during this era as well. Hybrid seeds became an important source of crop enhancement, and by mid-century 75 percent of all corn was hybrid, as was virtually all commercially grown wheat. The trend toward hybridization and high-yield agricultural practices, including monocropping, would eventually lead to the end of plant diversity in agriculture. In *Seeds of Change*, Kenny Ausubel states that "of the cornucopia of reliable, cultivated food plants available to our grandparents in 1900, today 97 percent are gone. Since the arrival of Columbus, 75 percent of native food plants have disappeared in the Americas."[21]

Artificial insemination breeding programs for cattle would reduce the number of commercially viable species from 274 in the late 1800s to less than 12 by the end of the century. Another chemical developed by Monsanto as a war weapon, 2,4-D, a defoliant also known as "Agent Orange," was the first selective herbicide and went on to become the most widely used weed killer in history. By the end of the decade preservatives known as BHA and BHT, both originally invented to keep K rations fresh during the war, were commercially introduced to extend shelf life in processed foods, and U.S. agricultural chemical production reached nearly 2 billion pounds.

World War II took women out of their homes and into the workforce, which put more money in the marketplace and instigated a need for convenience foods, as women would have less time to spend in the kitchen. By 1940 refrigeration had become an important part of American living, but it was not until the early part of the decade that home freezers began to have a commercial impact. It was around that time that the TV dinner made its way into our lives. Just twenty years after women were granted the right to vote, they were catapulted from their homes into the working world. Struggling to balance work and family, they wholeheartedly embraced new convenience technologies. It did not take long for us as a country to become obsessed with all things new. Technological advances were assumed to be in our best interest, proven or not. Faster was always better, higher production and profits foremost, consolidation the norm, and success synonymous with excess. Agriculturally this meant higher levels of production with a new quality standard. For some farmers a better-quality product meant a variety that was easy to grow; others wanted high yields; and still others looked for items that were easily transported. In the early 1950s flavor began to take a backseat to appearance and convenience.

In 1950 the Korean War again brought about the need for increased agricultural productivity, yet at the same time drought was devastating much of the United States, particularly in the Midwest, where sections of thirteen states were declared disaster areas. In 1954, in an effort to provide world hunger relief, give U.S. farm prices a boost, and protect U.S. farmers from world market collapse, the Agricultural Trade and Development Act was signed into law by President Eisenhower. This law empowered the USDA to buy agricultural surpluses and donate them to or trade them with foreign powers. The very same year, however, Eisenhower announced his disdain for agricultural price supports, blaming them for overproduction and surpluses. At that time, he spoke out in favor of holding surplus production as emergency reserves. Two years later, at the suggestion of the president, Congress voted to authorize a program that would reimburse farmers who voluntarily removed land from production. This action only served to reinforce a pattern of land abuse by farmers followed by quick-fix government subsidies for temporary soil replenishment.

The interstate highway system, funded by the Federal Highway Act, came to include some 50,000 miles of roadway connecting all major urban areas of the country; for the agriculture industry, it would eventually overshadow the importance of rail and waterways. In spite of diminished importance as a national means of transport, however, waterways gained import globally. In the twentieth century container ships and tankers began plying the seas, and at the close of the century shipping had become the most important form of transportation for the agricultural industry's global market. These advances in land and water transport, combined with improvements in refrigeration, have persisted in broadening the scope of the agricultural industry. An emphasis on economies of scale and increased efficiency has, in recent years, served to reinforce consolidation of industry, and small family farms are increasingly bought up by large-scale producers intent on industrializing agriculture. This is the single biggest threat to global plant biodiversity.

By 1960 farmers made up only 7 percent of the U.S. population. Many experts believed the world was on the verge of a global food crisis; yet Americans were spending less than 18 percent of their incomes on food, the lowest percent in history, and the lowest percent in the world.[22] A President's Advisory Council warned of world food shortages, and in the United States, widespread hunger and malnutrition became evident in rural populations. Over 15,000 farmers received subsidies, averaging under $25,000 per year, though some were given as much as $500,000 to $4 million. In spite of a dwindling number of farms, production actually rose. In 1963 the average farmworker was able to produce enough food and fiber for thirty-one people, but by 1973 that rose to fifty people.[23] At the close of the 1960s, American hunger and malnutrition was a major concern of governmental policy makers. Legislation was enacted to assist both hungry Americans and farmers who were still producing surpluses. Allocations for food stamp programs reached over $3 billion, and school lunch programs exceeded $8 million. The latter part of the decade brought a renewed focus on food safety issues and inspired Congress to pass a new "Truth in Packaging" law. The U.S. Meat Inspection Act took effect, and after a campaign by consumer advocate Ralph Nader the U.S. Food and Drug Administration (FDA) took steps to crack down on chemical and drug usage in livestock production.

The government was continuing throughout this period to educate the American population on the importance of nutritionally balanced diets. The first United States food guide was developed in 1916 by the USDA and consisted of five food groups: milk and meat; cereals; vegetables and fruits; fats and fat foods; and sugars and sugar foods. By the 1940s the food guide listed ten food groups, including water and eggs. Vegetables and fruits were split into three individual groups—leafy green and yellow vegetables; citrus, tomato, and cabbage; and other vegetables and fruits. Ten food groups were difficult for consumers to remember, so these groups were trimmed to four food groups by the late 1950s. One of the most familiar food guides was the "Basic Four," which contained four food groups—milk, fruit and vegetable, bread and cereal, and meat—and was used for nearly twenty-five years. These food guides were eventually replaced by the USDA Food Pyramid and by alternative pyramids such as those devised by Oldways Preservation Exchange and Trust.

The 1970s saw continued regulation of chemicals in the agricultural sector as well as governmental acts targeted toward better nutrition and health for the consumer. The U.S. Environmental Protection Agency (EPA) banned DDT after Rachel Carson's *Silent Spring* inspired intense debate and heightened consumer awareness. DES (diethylstilbesterol), a synthetic estrogen used in the 1950s, 1960s, and early 1970s by the cattle industry to promote increased weight gain with "feed efficiency," was also taken off the market (in 1979), but not before a number of young girls were diagnosed with a rare form of vaginal cancer. Warnings by the secretary of agriculture that proposed restrictions on chemicals would have a deleterious effect on the agricultural industry were largely ignored. The School Lunch Act and the Special Supplemental Nutrition Program for Women, Infants and Children (WIC) were initiated during the 1970s, and the FDA standardized food labeling by requiring manufacturers to include nutritional data and U.S. recommended daily allowances (RDAs). The EPA went on to ban additional insecticides as the USDA made attempts to ban nitrites in meat; it seemed that a continuous war was being waged between the interests of corporate profit and public health.

By 1980 the U.S. agricultural system was in dire straits. Government programs that once again paid farmers not to plant crops cost taxpayers over $20 billion. On average over 100,000 farms a year were being pulled out of production. In 1985 farm subsidies topped $52 billion, and by 1995 the figure increased more than sixfold. By the early part of the decade, the start of a real reversal was evident; the government looked to cut the school lunch program, and the USDA responded by announcing that ketchup could be considered a vegetable. At the time, nutrition in America was poor, health care costs were skyrocketing, and there was increasing starvation worldwide. The greatest irony of this decade is that in 1985 our taxes funded $500 million in aid to starving people in Ethiopia while as a country we spent $50 billion on weight loss products.

At the turn of the millenium issues surrounding nutrition, chemicals, food safety, and bioengineering are hotly debated. Irradiation as an insecticide and a method of spoilage reduction has been proposed by the FDA and remains controversial although it has been used on many food products without public knowledge for quite some time. The USDA has approved the first genetically altered virus, $18 billion has been allocated for nutritional programs, and the beef grading system has been changed (some believe to its detriment) in an effort to market leaner beef to a more "health-conscious" consumer. The FDA continues to address truth-in-labeling issues, and recently declared that companies may no longer use subjective terms such as *fresh*, *low-fat*, and *healthy* unless they can prove that they have nutritional implications. In an extremely controversial decision, the FDA recently approved the use of BST (rBGH), a genetically engineered synthetic hormone for dairy cows that increases milk production. Overproduction of milk and the resulting low prices are already proving to be major challenges for many farmers. Farm subsidies are currently costing taxpayers over $10 billion dollars, and consolidation in the industry has 1 percent of the businesses producing 90 percent of the revenues. Bioengineering is coming into the public eye, and a proposed government bill on organic farming recently received over 275,000 primarily negative comments from the public—the most of any initiative in history.

By 1978 the total irrigated land in the United States had risen to over 50 million acres—roughly 5 percent of all agricultural land—up from 3.5 million in less than ninety years. Today, agriculture consumes about 84 percent of the water used in this country; rainfall is included in this figure, though if only groundwater sources are taken into account, agriculture ranks second after industry as the highest user of water. Throughout much of the country, water quality is inextricably linked with agricultural practices. We see the results of increased water consumption in the form of altered watersheds, oversalinated soil, and a decrease in crop production, and we are also discovering that most of our waterways have been horribly polluted. It is not uncommon for pesticides to be found in drinking water supplies in farm states. In fact, the Clean Water Act was passed in 1972 because the EPA found that nearly 70 percent of river "impairment" was caused by agricultural pollution. More recently, the U.S. Geological Survey completed a study of 5,000 water samples from wells and waterways; at least one pesticide was present in half of all wells and virtually every river and stream tested. Health risks have increased, and taxpayers' dollars are being used to fund huge, expensive water filtration plants. Once again, the government has stepped in to help conserve this precious resource. In 1998 the USDA helped to fund a program that pays dairy farmers to help protect New York City's water supply, and the same type of initiative has been approved for the Chesapeake Bay, and the Illinois and Minnesota Rivers.

By 1994 only 1.5 percent of the U.S. population remained farmers. Currently an American farmer produces enough food and fiber for 128 people,— 96 Americans and 32 citizens of other countries. At the beginning of human history it took 5,000 acres to feed each member of a family. Today, in this country, four to five people can be fed for an entire year off one acre of land. Beautiful, flavorful foods have been bred for efficiency, appearance, and profitability. As rapidly as biodiversity shrinks, the processed foods industry expands. At the end of the twentieth century, the fat substitute Olestra has become entrenched in the marketplace, 60 percent of all soybeans are derived from bioengineered stock, and cancer researchers believe that over one-third of all future cancers will be diet-related—roughly the same proportion attributable to smoking. We now spend an average of 14 percent of our income on food, less than the people of any other nation. At the same time, we also spend billions on nutritional aids, vitamins, dietary supplements, and skyrocketing health care costs. One has to wonder how it came to be that we now spend more on diets and dietary supplements than we do on vegetables.

By the 1990s the world's population was rising by 80 million people a year, and by the year 2000 total global population had exceeded 6 billion. Some predict it will be 8 to 12 billion by 2050. If that sounds like tremendous growth, it is. In the first ten thousand years of our planet's agrarian history, the population grew to approximately 297 million, or just about 30,000 people per year. In the subsequent 2,000 years it has grown by 5.7 billion, or 2,850,000 per year. Every ten seconds the world's population increases by twenty-seven people, according to the Population Reference Bureau. The day you read this approximately 250,000 people will be born. Who will feed them? Will it be safe? What will it cost? These are the questions that lead us into the current debate about sustainability. The following chapter explores and defines the issues as well as possible solutions, both historically and through interviews with farmers, chefs, educators, and government officials.

Corn: An American Life

1540	Hernando de Soto and his soldiers find Muskogean ancestors raising corn in the Southeast.
1590	Thomas Harriot's *Brief and True Report of the New Found Land of Virginia* describes neatly planted cornfields. Corn grows for the first time in the British colonies.
1620	Governor Bradford reports the accidental discovery of corn by pilgrims.
1779	Richard Bagnal, an officer with General John Sullivan, finds sweet corn along the Susquehanna River. He plants the sweet corn in Plymouth, but it will not be popular with Americans for another seventy years and will remain largely unknown in the rest of the world for about another century after that.
1789	Baptist minister Elijah Craig distills the first bourbon whiskey in Kentucky County (later known as Bourbon County). Craig's corn whiskey will become more popular than rum or brandy in America.
1815	March 21. Thomas Jefferson writes, "We have had a method of planting corn suggested by a Mr. Hall, which dispenses with the plough entirely." "Check-row" corn planting, already widely practiced by Native Americans, has been patented by Mr. Hall.
1820	The mechanical cultivator is introduced, and hand weeding of corn is no longer necessary.
1850	Milo maize (kaffir corn) is cultivated as livestock feed.
1857	Martin Robbins, a Cincinnati inventor, introduces the first check-row corn planter.
1860	The U.S. corn crop reaches 839 million bushels.
1864	Illinois inventors John Thompson and John Ramsay patent a check-row planter. Canned corn is reported to be a staple food in Colorado. The Pennsylvania Dutch dry corn for the first time.
1870	The U.S. corn crop tops 1 billion bushels for the first time.
1871	Corn exports reach 1 million bushels.
1872	Farmers reap a bumper corn crop; surplus corn is fed to cattle, leading to the start of stock feeder cattle in the Midwest.
1875	A machine is introduced that strips corn from the cob, making widespread canning of sweet corn possible.
1880	Corn production reaches 1.5 billion bushels, and exports have increased to 116 million bushels. Prices take a 15 percent nosedive, and many farmers cannot afford to ship their product to market.

Corn: An American Life (continued)

1880   Production reaches 2 billion bushels, most of which goes to hog and cattle feed.

        Aunt Jemima's self-rising pancake mix, which uses corn as a main ingredient, is introduced.

1898   Sanitas cornflakes are introduced by J. H. Kellogg and his brother Will Keith. The world's first cornflakes unfortunately turn rancid on store shelves.

1901   The steam corn sheller goes through a bushel of corn nearly seventy times faster than the same job done by hand.

1906   Corn production reaches 3 billion bushels.

1911   Battle Creek, Michigan, plants produce cornflakes under 108 different brand names. Mazola Corn Oil is introduced for home consumption.

1916   The first hybrid seed corn is purchased by a farmer in Jacobsburg, Ohio. Average corn production is 25 bushels per acre.

1921   Average corn production is 28.4 bushels per acre.

1924   Farmer/editor Henry Wallace wins the Iowa Corn Yield Contest (his own idea) with hybrid Copper Cross.

1926   Wallace founds Pioneer Hi-Bred.

1930   Favorable yields from hybrid corn piques the interest of corn belt farmers, who were initially very resistant to it.

1932   Frito's Corn Chips are introduced.

        James Wallace (Henry's brother) predicts hybrid corn will overtake future markets and begins his own company. He will eventually sell 645 million pounds of corn seed a year.

1933   The Emergency Farm Bill reinstates 1909–14 average prices on several staple crops, including corn.

        Golden Cross Bantam is the first hybrid to be planted on a large scale.

1935   Pfister Hybrid Corn begins with a high-yielding double-crossed hybrid.

1944   Average corn production is 33.2 bushels per acre.

1948   Seventy-five percent of U.S. corn is hybrid.

1959   It is necessary for one corn farmer to have at least 1,000 acres in order to be viable.

        There are 1 million U.S. farms under 50 acres.

1960   Average corn production is over 40 bushels per acre.

1961   Corn surplus exceeds 2 billion bushels, a record that will go untouched for more than twenty years. In response, President Kennedy introduces an emergency feed grain program to reduce production acreage.

Corn: An American Life (continued)

1963   The U.S. corn crop exceeds 5 billion bushels.

1965   Some U.S. farmers report yields of up to 120 bushels per acre.

Total U.S. production is at 9 billion bushels per year.

Eighty-five percent of corn produced is used as animal feed.

1967   High-fructose corn syrup (HFCS) is developed by Clinton Corn Processing Company. Archer Daniels Midland (ADM) and several other companies rush to build HFCS plants.

1968   U.S. farmers own a total of 660,000 mechanical corn pickers and shellers.

The average U.S. farm gets 70 bushels per acre of corn, although some report as much as 200 bushels per acre.

1968   Opaque-2, a high-protein corn that has excellent potential for putting an end to Third World malnutrition problems, is developed.

1970   Orville Reddenbacher introduces his Gourmet Popping Corn.

1972   Average per capita consumption of high-fructose corn syrup is 1 pound.

1973   Average corn production is at 96.9 bushels per acre.

1974   High-fructose corn syrup is recognized as a cheap substitute for cane or beet sugar, and production skyrockets. Average consumption of HFCS is 10 pounds.

1980   Drought hits the corn and cotton belts hard, and conditions rival those of 1934, 1936, and 1953. Winter wheat is harvested before the drought, but feed grains suffer—the corn crop alone drops 14 percent, to 6.6 billion bushels.

1994   U.S. farmers report record highs in grain crops, including corn, yet the number of farms has dropped to 1.9 million, its lowest point since 1850.

1997   Over 6 million acres of Yieldgard corn, a variety bioengineered by Monsanto, are being grown.

1999   Archer Daniles Midland demands that bio-engineerd products (including corn) be separated from nonbioengineered products.

▪ 2 ▪
Sustainability

> I am pessimistic about the human race because it is too ingenious for its own good. Our approach to nature is to beat it into submission. We would stand a better chance of survival if we accommodated ourselves to this planet and viewed it appreciatively instead of skeptically and dictatorially.
>
> —excerpted from *Silent Spring*, Rachel Carson

Mention the word *organic* to most Americans, and their eyes begin to glaze over. Many people I have talked to believe that the organic movement is nothing more than media hype—just another ploy to convince us to pay more for our food. Most people simply do not realize how many carcinogenic chemicals are utilized in conventional agriculture. Sustainability, a separate but related issue, is a concept much further removed from the American consciousness. In this country, because our food supply is seemingly endless and plentiful, we have a difficult time believing that it may be unable to sustain us indefinitely. Today's conventional farming consistently overworks the soil and pollutes our waterways. It relies heavily on certain breeds and varieties of plants and animals that may, for instance, produce higher yields, or are easier to transport over long distances, causing a progressive narrowing of biodiversity on our nation's farms. Nutritional value and flavor are frequently sacrificed in favor of the bottom line. Overall, Americans lack the information they need to make healthy food choices. Rather than take the time to educate ourselves, we seem to prefer to ignore the problems created by our current methods of food production. Additionally, we have grown accustomed to getting exactly the foods we want year-round, and farmers around the world are happy to meet our market demand; as a nation, we can afford it. Our planet cannot.

Sustainability is a complex issue. It embraces both the continued sustenance of the world's people and the long-term protection and rejuvenation of our natural resources. Without higher levels of production, the world's people will starve. Those who seek to increase production from the same

amount of land using conventional farming techniques are generally proponents of increased chemical inputs and continued mechanization.
Protectors of the environment are pushing for a return to local, regional,
and seasonal food supplies, a reduction in chemical application, and the
research and development of alternative methods to increase yields. They
know that without clean water and healthy soil we cannot produce a viable
food supply. They also understand that if we continue to transport foods
across vast distances to satisfy our yen for out-of-season products, we will
deplete our supplies of fossil fuels and cause infinitely greater damage to
the air we breathe.

What emerges are two divergent agricultural philosophies—industrial
and sustainable. Industrial agriculture, also called agribusiness, is large-scale
farming. Its hallmarks are massive beef and pork feedlots, poultry barns
where as many as 20,000 chickens are being raised, and produce farms where
mile upon mile of vegetation is mechanically harvested. In general, agribusiness farms focus on one product, and in some cases the ground is even
sterilized and chemically treated to keep other plant life out of the fields.
The business model used in industrial agriculture focuses on producing
maximum product yield through the standardization of production techniques combined with intensive mechanization and continuous technological
upgrades. The other philosophy, often referred to as "alternative," is known
as sustainable agriculture. It is generally regarded as a method of small-
scale farming that requires lower capital investments but is much more
labor-intensive than its counterpart. Sustainable agriculture uses a business model that measures the quality of production rather than production
volume. While sustainable farms can generate profit, their business philosophy places a high value on the earth and its inhabitants.

Because no easy solution exists, it is our social responsibility to understand how food moves from the farm to our table. The choices we make in
our communities will determine the balance between industrial agriculture
and sustainable agriculture. Many commercial food producers would argue
that today's industrial agriculture is more sustainable than any other type
of farming, past or present. We are, after all, feeding more people using less
land than at any other time in our history. But how dependent do we want
to become on an industry that relies so heavily on chemicals and controversial breeding techniques? It is easy to react viscerally to either or both
philosophies. Chemicals cause cancer. Our water supply is in danger. Our
soil is eroding at alarming rates. Urbanization is consuming available farmland with incredible celerity. What is the future of our planet?

Fortunately, quite a few nascent movements are currently working to
protect our soil, water, and air, and a large number of businesses and organizations are focused on bringing sustainability issues to the fore. Groups
like Chefs Collaborative 2000 and Mothers & Others for a Livable Planet
are encouraging home cooks and restaurant chefs to employ proactive mea-

sures in their kitchens. Educational institutions such as the Culinary Institute of America and the New England Culinary Institute are teaching a new generation of chefs the importance of seasonal and regional menus, and small farmers are actively looking for more creative ways to stay in business in spite of the push toward industrial agriculture. Ultimately, though, the consumer will be the catalyst for change, and education is the first step.

ORGANIC AND SUSTAINABLE: HOW ARE THEY RELATED, AND WHAT DOES IT MEAN?

Organic farmers grow a variety of crops, practice crop rotation, and alternate fields with cover crops that naturally replenish the soil's nitrogen levels. Weeds are removed by hand or machine (on a more frequent basis than conventional agriculture, because of the lack of sterilizing chemical applications), and insects are controlled by the introduction of naturally occurring mating disrupters, other "beneficial" insects, or organic insecticides. Anything applied to the soil or crop adds to the biological diversity and ecological health of the farm environment.

Large, conventional agribusiness farms often produce a single specific crop, year after year. Monocropping, as it is known, is preferred because specialized equipment can replace field-workers, who move more slowly and ultimately represent a greater financial burden to the business. Monocropping saves time and money. However, the consequences of monocropping include stripping the soil of essential nutrients, the intensive use of chemical herbicides, pesticides, and fertilizers, and the production of food for ease of handling without much consideration given to flavor or nutritional content. Once the soil is exhausted, farmers rely on chemicals to maintain crop production. The soil is rejuvenated with manmade fertilizers, and weeds and insects are controlled with herbicides and pesticides, many of which are known by the EPA, FDA, and USDA to have an undesirable impact on air, soil, and water quality.

In the late 1960s and throughout the 1970s American society was in philosophical turmoil. Consumers questioned the government's motives on everything from our participation in the Vietnam War to our agricultural practices. During this time, there was a movement away from mass production to sustainability, and consumer demand for unadulterated, or organic, products increased year after year. Organic farmers requested some type of business blueprint for the industry. In 1980 the USDA produced its *Report and Recommendations on Organic Farming*, commissioned by then secretary of agriculture Bob Bergland, which included case studies of sixty-nine organic farms. The report solidified the sustainable agriculture movement, and was the first government publication to acknowledge and discuss an alternative to industrial agriculture.

Chefs Collaborative 2000

CC2000 promotes sustainable cuisine by teaching children, supporting local farmers, educating one another, and inspiring the public to choose good, clean food. Founded in 1993 by Oldways Preservation and Exchange Trust and a group of leading chefs, the Collaborative is a national nonprofit membership organization of chefs who are dedicated to the ethic of sustainable cuisine. The following is the CC2000 mission statement:

> We, the undersigned, acknowledging our leadership in the celebration of the pleasures of food, and recognizing the impact of food choices on our collective personal health, on the vitality of cultures and on the integrity of the global environment, affirm the following principles:

Statement of Principles

Food is fundamental to life. It nourishes us in body and soul, and the sharing of food immeasurably enriches our sense of community.

Good, safe, wholesome food is a basic human right.

Society has the obligation to make good, pure food affordable and accessible to all.

Good food begins with unpolluted air, land, and water, environmentally sustainable farming and fishing, and humane animal husbandry.

Sound food choices emphasize locally grown, seasonally fresh, and whole or minimally processed ingredients.

Cultural and biological diversity is essential for the health of the planet and its inhabitants. Preserving and revitalizing sustainable food and agricultural traditions strengthen that diversity.

The healthy, traditional diets of many cultures offer abundant evidence that fruits, vegetables, beans, breads, and grains are the foundation of good diets.

As part of their education, our children deserve to be taught basic cooking skills and to learn the impact of their food choices on themselves, on their culture, and on their environment.

NATIONAL ORGANIC STANDARDS

Throughout the decade following the first governmental report on organic agriculture, the industry grew at an unprecedented rate. In 1997 Katherine Dimatteo, executive director of the Organic Trade Association (OTA), reported that "organic sales have increased 20 to 24 percent each year" over the previous three or four years.[1] In fact, sales of organic products rose from $1 billion in 1990 to over $4 billion in 1997. Today organic production continues to expand, and sales are predicted to top $6.5 billion in 2000. This unprecedented growth did not escape the notice of conventional farmers in the early 1990s. Some began to label their products "organic" even when they were not. Seeking protection, true organic farmers began lobbying for national organic standards.

By 1990, eleven states and thirty-three private agencies had adopted local standards, and in 1999 fourteen states—Colorado, Idaho, Iowa, Kentucky, Louisiana, Maryland, New Hampshire, New Mexico, Nevada, Oklahoma, Rhode Island, Texas, Virginia, and Washington—had state-run certification programs. The specific standards and regulations imposed by these agencies grew out of local demand, however, and there was no national set of regulations that guaranteed food quality standards and specified allowable production techniques from one geographic area to the next. Confusion generated by the industry's inconsistent standards did little to benefit consumers or producers, so organic farmers and processors looked to the federal government for help in defining and standardizing a set of rules for their growing industry.

The Organic Foods Production Act of 1990 proposed to establish a National Organic Program, which would include regulations for organic production and handling of agricultural products, as well as a national list of approved synthetic substances. In addition, the National Organic Program was enlisted to develop a certification program available to all operations wishing to meet the requirements for organic labeling. Over the following six years the government-appointed National Organic Standards Board (NOSB), comprised of a cross-section of industry experts, polled organic communities around the country and held public input hearings as a means of gathering information. Finally, in early 1997, after painstakingly careful consideration, the NOSB presented the U.S. Department of Agriculture (USDA) with its recommendations for national organic standards. These included a definition of the term *organic*, organic labeling standards, the use of antibiotics in organic livestock production, and the use of pesticides, herbicides, and fertilizers in organic food production.

In December 1997 the USDA released a 300-page document describing the proposed rule for the National Organic Program. It departed shockingly from the content and spirit of the NOSB recommendations, ignoring what most considered to be the very essence of organic farming. An outraged public responded venomously. Among others, Chefs Collaborative 2000 (CC2000) came out against the proposal, as did Mothers & Others for a Livable Planet, Oldways, the Northeast Organic Farming Association (NOFA), and all of the state-run organic farming organizations. In early 1998 the Vermont chapter of CC2000 sent a letter of protest to its members and the USDA. The following is an excerpt:

> We, the professional chefs and restaurateurs of this state, believe that the
> United States Department of Agriculture's (USDA) proposed rule for the
> National Organic Program unwisely dilutes and blurs established regional,

state and international distinctions between "organic" and non-organic foods
and serves only to confuse the issue for the public.

The Collaborative and scores of others in opposition to the guidelines were
specifically united against including the "Big Three"—genetic engineering,
fertilizing with municipal sewage sludge, and irradiation—as allowable
under organic labeling regulations. Additionally, many groups and individuals took issue with the use of antibiotics, nonorganic feed, and intensive
confinement for animals raised for meat and dairy production. The proposed guidelines also specified that organic producers would not be permitted
to inform customers as to the absence of questionable ingredients (such as
rBGH) in their products. The chefs of the Vermont chapter of CC2000
expressed their outrage at the close of their letter, stating,

> Finally, we find it nothing less than incredulous that the USDA dismissed,
> in significant measure, five years of public input to the National Organic
> Standards Board (NOSB), whose recommendations are far more aligned
> with a generally accepted understanding of what is meant by organic and
> which, had they been accepted, would have gone a long way in resolving
> these issues.[2]

This letter, along with thousands of others like it, piled up at the USDA in
the months following the proposal's release. In fact, the Department of Agriculture received over 275,000 mostly negative responses, a number
unprecedented in the history of the agency.

The public outcry was so deafening that the awestruck USDA was forced
to table the proposal and resubmit it for revision. One staff member con-
ceded, "We underestimated the strength of commitment to the term *organic*
that exists out there." Agriculture Secretary Dan Glickman has stated that
the USDA will now make "fundamental" changes to the proposal, promis-
ing to jettison genetic engineering, municipal sludge, and irradiation. The
question to be asked, of course, is how, in the face of the NOSB's tireless
research and the desires of organic producers themselves, could the USDA
have come up with a plan so far off the mark? As with far too many political issues, the answer may come down to money. Reed Karaim of the
Washington Post cast some light on the issue when he reported that the
USDA and the entire U.S. government has a financial and political interest
in convincing us and the rest of the world that conventional farming is the
best alternative. Karain wrote,

> Organic farming is inevitably seen as an implicit criticism of [conventional
> agriculture]. What's more, each of the controversial items that USDA has
> allowed into the organic standards has its defenders. The biotech industry
> insists that genetically manipulated products are as natural as any. The food

processing industry and the Food and Drug Administration support the use of irradiation. The Environmental Protection Agency likes the idea of sewage—they prefer the term "bio-solids"—being recycled through agricultural use. Animal confinement and the use of hormones are supported by agribusiness and producer groups.[3]

Indeed, the influences described by Karain can be seen in the portion of the proposal that deals with labeling. According to the proposed rules, the word *organic* can be used on labels, but all other illustrative commentary is forbidden. This means, for instance, that farmers would not be permitted to mention that their animals are raised antibiotic-free or that their potatoes are not genetically engineered. These may seem like minor points, but if the guidelines were passed as written, a plant grown from bioengineered seeds, fertilized with municipal waste, and then irradiated could be deemed organic, and the public would be none the wiser. A farmer growing a truly "organic" product would not be allowed to adequately inform his customers by including phrases like "not irradiated" or "not bioengineered" in labeling. The USDA argued that this rule would eliminate unfair competition between established organic producers and newly certified, or "transition," organic farmers, and could lead the public to think that bioengineering, irradiation, and municipal sludge are undesirable.

The bottom line is that agribusiness is a well-funded constituency, and the USDA must be very sensitive to its relationship with it. Companies such as Monsanto are responsible for billions of dollars of bioengineered products, and the USDA and the FDA, among other government agencies (who, as Karain pointed out, support the use of irradiation and municipal sludge), cannot afford to endanger their relationship with agribusiness. Instead, these governmental entities would give the least benefit to the very people who are responsible for the success of the organic industry in the first place.

Among the U.S. senators active in the organic and sustainable agriculture movement, Senator Patrick Leahy of Vermont demonstrates a genuine understanding of the issues. For example, he authored the 1990 Farm Bill, considered the "greenest" farm bill of the twentieth century in that it challenges the government's assertion that the only answer to food production in this country is large-scale technology-based farming. In addition, his Farms for the Future program is credited for saving 200 farms from bankruptcy. With a touch of irony, Leahy muses about agribusiness in the late 1990s: "For years I tried to get the Organic Standards Act through and ... was blocked by the mega-producers. Their lobbyists are all over the Hill, and they were acting as though I had threatened to cut off the wheat fields of Kansas and Oklahoma! Then suddenly they realized this is a multi-billion-dollar industry, and they want to come in and take it over themselves."[4] While Leahy continues to be the voice of the movement, the USDA has yet to come up with a standard definition of the term *organic*. As a consequence, agribusiness is free to keep its labeling vague, at the expense of the American public's right to know the truth about their food choices.

It is not the presence of agribusiness and conventional farmers or their eagerness to partake of the profits generated by the organic movement that really bothers me. After all, if there is a defined set of organic standards enforced by the appropriate government agency, then theoretically, big business and small organic farmers will be working with the same rules, and consumers will feel justifiable confidence in their purchases. In fact, if big business adopts and adheres to organic standards, the consumer ultimately comes out ahead; with greater production, prices are likely to fall. This is where the debate about organics and sustainability takes a disarming turn. It is easy to preach the benefits of organic farming—that is, until you look at it on a large scale and realize that it is possible to farm organically and not sustainably. In an article in *Organic Farmer: The Digest of Sustainable Agriculture*, Fred Kirschenmann illustrates this point well:

> An "organic" monocropping carrot producer on the West Coast who purchases botanicals from an East Coast vendor to control his pests and imports fish emulsion and seaweed from Japan for his fertilizer and then sells his crop to an outlet in Massachusetts is probably not sustainable. The producer who qualifies as organic simply because he does not use any "unacceptable" materials to produce his crops, but who does nothing to recycle nutrients or to create a healthy, diverse growing environment to naturally control pests, will find himself mining the soil and wasting resources in a manner that cannot be sustained. Organic agriculture that only seeks to avoid using "unacceptable" materials to qualify for the safe food market cannot be sustained over the long term and does not deserve the label "organic."[5]

In contemplating Kirschenmann's assertion that organic farming that is not sustainable should not be labeled organic, I was reminded of a presentation given by Joan Gussow, in 1996 at a CC2000 retreat. Gussow is one of the most inspirational people I have ever had the good fortune to give ear to on the subject of organic and sustainable food choices. Professor emeritus at Teachers College, Columbia University, Joan is the author of a number of books, including *Chicken Little* and *Tomato Sauce and Agriculture: Who Will Produce Tomorrow's Food?*, and a passionate proponent of the ideal of the "relocalization of the food supply." During the 1996 CC2000 retreat, Joan spoke to the audience about her experience during the discussion period at an organic processing workshop, in which the question was posed, "What restrictions should be put on the use of substances not available in an organic form but 'essential' in the production of specific processed products?" While speaking to us at the retreat, Joan remembered the conversation as extremely unnerving. She found the entire concept of nonorganic organic very distressing. It was her opinion at the time that there were some products she preferred not to see in organic form, "an organic gummi bear, for example." Joan asked her group, "[Is] the goal to have a parallel organic food supply, one with organic Twinkies, organic Eggo Toaster Waffles ... and organic Count Chocula Cereal?"[6] I wondered how the Twinkie, rumored among certain generations of schoolchildren to be the only food that could withstand a nuclear holocaust, could be organic. Joan's organic Twinkie was what turned me on to sustainability, and it is the concept to which I have returned throughout my research. Is the organic Twinkie possible? Do we even want an organic Twinkie? Is it possible to continue to support the world's population if we eschew the concept of the parallel food supply?

THE SUSTAINABLE PLANET

The Land and Soil

The United States is blessed with rich soil for all types of farming. However, throughout our history, landowners have focused on profit rather than sustainability. Even Thomas Jefferson, often seen as a romantic visionary for the American farm, was guilty of putting profit before the health and vitality of the land. Jefferson was a tobacco farmer. He had his slaves clear vast quantities of land for cultivating the lucrative crop, which wreaks environmental havoc on the land. Jefferson is not alone. The vast majority of farmers have consistently worked the land in a way that causes the erosion of good, healthy, vital soil in exchange for short-term profitability.

Conventional farming depletes topsoil by an average of 3 billion tons annually. The Soil Conservation Service estimates that every year we lose an average of one-sixteenth of an inch of soil per acre. In sixteen years we lose one inch of soil. In 100 years, we lose six inches! Estimates vary as to how long it takes for one inch of topsoil to be created, but under natural conditions, they range from 300 to 1,000 years. In agricultural systems, when fertilizers replace plant nutrients and when organic matter is allowed to accumulate, one inch of topsoil will be rejuvenated in thirty years.[7] According to these calculations, the average net loss is one inch every thirty years; two inches over the life span of an average human generation. There are groups, organizations, and people working together and individually, to preserve and regenerate remaining topsoil. Numerous states have soil conservation agencies—Alabama, Iowa, and Virginia, for example—and both the USDA and FSA have programs that work on these issues. Additionally, even though most conventional farmers continue to use the straight-row method, they have adopted low- or no-till practices, which leave the roots and dead stems of the previous year's crop to hold the soil in place.

Few farmers, however, are as openly passionate about their dedication to the soil as Michael Ableman. Author, photographer, longtime sustainable agriculture activist, and owner of Fairview Gardens in Goleta, California,

Ableman is among the most extraordinary people I encountered during my research for this book. His farm, located less than 100 miles north of Los Angeles, is a tiny oasis in the midst of seemingly endless suburban sprawl. To get to Fairview Gardens, where I spent the better part of a morning with Ableman, I drove through the characteristic gray haze that extends from Los Angeles via a network of highways, past mile after mile of tract housing and strip malls. In the face of that extreme urbanization it was difficult to visualize the fertile, agricultural area that once existed there. As I neared the farm I drove past a library and up a long driveway, the top of which opened onto a glorious reminder of the past. Fairview Gardens is a place of lush growth and variegated greenery. The air there is perfumed with the unmistakable scent of ripening produce commingled with the aroma of sun-warmed earth. Once part of a vast 4,000-acre parcel, this celebrated "urban farm" now covers only twelve acres and stalwartly holds its ground against the intrusion on all sides, of an advancing metropolis.

Ableman is an impassioned organic farmer, teacher, and advocate of "good food" and sound agricultural practices "that work with nature rather than against her." Above all, he is a dedicated keeper of the earth. His unbridled enthusiasm is, at the very least, contagious. At his best, Ableman changes the way people think about their food and the earth forever. His views on soil and its connection to our food supply made an irrevocable impression on me. As we talked, I could imagine him bending to retrieve a handful of soil as he explained his method of teaching children on the farm. He begins his lesson with an earthward gaze. He says, "This is where our food comes from." Simple, but not necessarily apparent to a generation of youngsters who believe that food somehow magically appears on supermarket shelves, wrapped securely in plastic. He asks them to "examine the soil closely and smell it deeply," explaining that "one pinch of living soil contains millions of forms of life, and the strength and well-being of all living things is directly ... tied to the health of the soil." Abelman's agricultural philosophy—that everything in life begins and ends with healthy soil—evolved as his proficiency as a farmer grew. At the beginning of his career he found himself continually blaming external influences for problems with his crops, but over time he came to understand that "the fertility and condition of the soil" is of utmost importance. He told me that after all these years of work he truly "understands why some cultures call [soil] the earth's placenta."[8] This simple understanding dictates everything he does.

Water

Without clean water, the human body, which is 70 percent water, is at risk for disease. Without clean water, farmers cannot produce wholesome food. The relationship between water and crop production is a critical connection for our good health. According to the U.S. Water Council, 80 percent of water from all sources is used by the agriculture industry. American farmers harvest approximately 325 million acres of crops a year. At this production rate, the United States faces a definitive threat to its water supply within the next twenty years. Water used in the irrigation process will come under increasing scrutiny.

Nearly all water contains some sodium chloride. During the irrigation process this naturally slightly salted water is continuously sprayed over the land and crops. When it evaporates, a salty residue is left behind. Each time the land is irrigated the amount of residue is compounded, eventually causing a marked increase in the overall salinity of the soil. In *New Roots for Agriculture* Wes Jackson delineates the future cost:

> Seven hundred thousand tons of salt are added to the Colorado [River] each year from the Grand Valley of western Colorado alone. The consequence is an economic impact of more than $50 million each year in reduced crop yield and corroding of equipment. This cost could easily double in the next 20 years, even at a fixed dollar cost. Salinity at the headwaters of the Colorado system is less than fifty milligrams per liter. At the Imperial Dam near Mexico, salinity has increased eighteen-fold to 900 milligrams per liter of water. The concentration is expected to reach 1160 milligrams/liter by the year 2000. The economic losses for each one milligram per liter addition is over $200,000.[9]

If our society continues to rely on conventional agriculture, with its dependence on irrigation, pesticides, and fertilizers, we must face the reality that little clean water exists for our survival. The EPA has already found ninety-eight pesticides (including DDT) in the groundwater of thirty-eight states. How will we produce enough food to sustain our crops, much less fulfill our own biological need for clean drinking water? Where will the water come from? These are the questions we must ask of agribusiness and our government.

A number of organizations, both regional and national, are currently working toward a solution to the oversalinization problem. In fact, the U.S. Salinity Laboratory (USSL), an arm of the USDA, has been working since 1947 "to develop, through research, new knowledge and technology dedicated to the solution of problems of crop production on salt-affected lands, sustainability of irrigated agriculture, and degradation of surface- and ground-water resources by salts, toxic-elements, and pesticides."[10] The USSL has also published a handbook addressing the issue, which is available to both farmers and consumers. Additionally, author and Ecology Action founder John Jeavons's biointensive farming methods and other no-till practices focus on reduced water consumption and alternatives to irrigation. Continued efforts such as these will help avoid further soil degradation.

The Watershed Agricultural Council, a farmer-led organization created to oversee a project that protects the New York City water supply, is already making progress in controlling agricultural practices to ensure potable water for city residents. Funded by the city of New York, the council's main focus is on the safekeeping of the Catskill-Delaware Reservoir, which supplies 90 percent of the city's water through a complicated system of underground pipes. The water travels over ninety miles before it reaches its final destination. Interestingly, Hilary Baum, marketing adviser to the Watershed Agricultural Council and coauthor of *Public Markets and Community Revitalization*, told me that the city has identified environmentally sound agriculture as the preferred use for land in the watershed; if properly executed, it can actually protect water supplies. Over 300 farms of varying types and sizes are currently participating in the program. It requires a commitment to an evaluation process that guarantees that farmers will do "whatever it takes," including manure management, reduction of chemical inputs, and the implementation of cow- and chemical-free natural growth areas along reservoir tributaries, to improve their operations.[11] With those systems in place, land will not be at risk for oversalinization, and the water supply will remain safe for city residents.

PLANTS AND BIODIVERSITY

In *Seeds of Change*, Gary Nabhan wrote, "We now lose to oblivion three plant species an hour. Of the cornucopia of reliable cultivated food plants available to our grandparents in 1900, today 97 percent are gone. Since the arrival of Columbus, 75 percent of native food plants have disappeared in the Americas."[12] As a chef, I am deeply troubled by the above quotation, for the more we move away from biodiversity, the more we put ourselves and our planet at risk.

Hybrids, which have their contemporary roots in Austrian botanist Gregor Johann Mendel's pea experiments in the nineteenth century, have a complex genetic makeup, a minimum of two distinct parents, and the inability to reproduce an identical crop from one generation to the next. The promise of higher yields, better portability, and increased profits brought about the widespread cultivation of hybrids and dramatically reduced the use of nonhybrid, or traditional, strains. The Food and Agriculture Organization (FAO) of the UN reports that traditional agriculture utilizes 80,000 species of plants, whereas industrial agriculture relies on just 150 varieties. These varieties are grown deliberately to withstand the rigors of shipping, a battery of heavy machinery and toxic chemical pesticides as well as for cosmetic appearance.

Not once is there a mention of flavor, nutrition, or safety.

Agribusiness systematically breeds out plant varieties it considers poor profit performers in favor of hybrids that turn a profit more efficiently and quickly. Research by the Rural Advancement Foundation International

(RAFI) indicates that "of the seventy-five kinds of vegetables traditionally grown in the United States, 97 percent of all the varieties have become extinct in less than eighty years." In other words, we have lost or destroyed virtually all varieties of every vegetable grown on American soil in less than a century. Fruit's fate has been the same. Between 1804 and 1905, there were over 7,000 apple varieties in the United States. Today, 6,121, or 86 percent, are extinct. These varieties are gone from our orchards because hybrids are more profitable for agribusiness. [13]

History, if left unchallenged, is sure to repeat itself. Today, even though we know that alternative business models exist, most major food crops are in danger because there are so few varieties under cultivation. Consider what would happen if a disease destroyed an entire crop. It is quite possible that we would face a worldwide food shortage. Most of our wheat, potatoes, peas, and oranges come from only a handful of varieties each. All the wheat grown in the United States is planted in only nine varieties; half the soybeans in six; 96 percent of peas in two; 90 percent of Florida oranges in only three; 66 percent of rice in just four varieties. Finally, four varieties of potatoes account for 75 percent of U.S. production. Have we learned nothing from history? It is unbelievable that the potato, of all vegetables, should be so perilously close to the edge. After all, every American child is taught in elementary school that the potato blight in Ireland—which would not have struck so disastrously if more varieties had been grown—caused the starvation of more than 1 million people.

Lack of biodiversity in farming is not limited to produce—in fact, animal husbandry may be at even greater risk. The white leghorn lays 90 percent of all chicken eggs in the United States, and 70 percent of the nation's dairy herd is Holstein. It is believed that over the next several decades there will be a 45 percent loss in North American livestock species. There are currently 175 recorded livestock breeds being raised in Canada and the United States, and 80 of them are listed by the American Minor Breeds Conservancy (AMBC) as "declining or rare breeds." It does not take an incredible leap of imagination to understand that egg production in this country would come to a virtual standstill if the white leghorn were stricken with disease. The same could happen to the dairy industry if the Holstein were to develop a new and untreatable ailment.

The American Livestock Breeds Conservancy is a nonprofit organization, founded in 1977, working to protect approximately 100 breeds of cattle, goats, horses, sheep, swine, and poultry from future extinction. The ALBC's programs include gene banks to preserve genetic material, rescues of threatened populations, research on breed characteristics, education about genetic diversity, and technical assistance to breeders and farmers. This type of organization may ultimately be the resource of our future supply of animal food products if the consolidation and specialization of agribusiness continues to cause a reduction in genetic diversity. If genetic diversity dwindles, and if a disease that is resistant to contemporary medicine finds its way into our cattle, heritage breeds may become the next generation of livestock.

The U.S. government operates a seed-saving program, whose goal it is to ensure a supply of seed for propagation in case of disease or natural disaster. The National Plant Germplasm System (NPGS), which has been in operation since the mid-1900s, holds seeds in repositories such as the National Seed Storage Library (NSSL) in Fort Collins, Colorado, and strives to "document, preserve, and maintain viable seed and seed propagules of diverse plant germplasm in long-term storage, to develop and evaluate procedures for determining seed quality of accessions as well as the distribution of seed for crop improvement.[14] Dubbed the "Fort Knox of Seeds," the NSSL is a building of concrete vaults, steel-plated ceilings, and high-tech waterproofing. The vaults house approximately 400,000 seed specimens and 34,000 samples of plant material necessary for grafting apples, grapes, and berries of all kinds. I recently spoke at length with Steve Eberhart, laboratory director of NSSL in Fort Collins. He feels confident that the program is working hard to protect the country's germplasm, but explained that the types of material housed at Fort Collins are, for the most part, commercially grown types, and that heirlooms, or "non-products" (not commercially viable) materials, are left to the "seed savers" to preserve (see below).

Some people may dismiss the fear of animal and plant loss, believing that predictive science and preventative medicine will ultimately save us from our own course of destructive behavior. It could happen. If we are lucky we might still be left with bland milk, two types of peas, three varieties of oranges, and eggs from one chicken. As a chef whose business it is to create interesting dishes from flavorful foods, I am more than a little concerned about the growing shortage of traditional, unique, or unusual plant crops.

When we eradicate the essence of food through hybrids, we lose something natural on every level. The tomato, for example, is a ubiquitous and versatile food universal to world cuisine. However, if a tomato is picked at the peak of ripeness, it is impossible to transport it over a significant distance. The agribusiness solution comes through hybridization science. This tomato is tough-skinned, densely fleshed, and flavorless; but it travels well. For myself and other chefs throughout the country, this is not an acceptable trade-off. I might just as well cook with cardboard concassé. Recently, I have begun exploring heirloom varieties, which are typically cultivated on small farms and in backyards. Unfortunately for chefs around the country, those are exactly the treasures the NSSL is *not* storing. They are not as productive, not as easily transported, and not as profitable.

The Seed Savers Exchange (SSE), headed by Kent Whealy, is one of the organizations working to preserve our heirloom varieties. Founded in 1975, the exchange is a nonprofit organization with over 8,000 members who grow heirloom vegetables, fruits, flowers, and herbs for distribution among themselves and the public. Kent believes that the exchange has saved several thousand varieties of "living heirlooms" from extinction. The seed bank holds in excess of 18,000 vegetable varieties alone. Heritage Farm, the organization's headquarters, is a 170-acre living museum supporting and preserving 2,000 historic varieties of vegetables, 700 varieties of nineteenth-century apples, and 200 hardy grapes.

Kent explains that the Seed Savers Exchange is an important resource for two reasons; first, "for the genetic resources [because what we have at the Exchange] is essentially all of the material that we will ever have to breed the future's food crops. [The second] involves our history and the culture—these [food plants] were brought from all over the world." The ethnic diversity of this country has become the culinary and cultural diversity of American cuisine. If we lose the plants that are directly tied to the historical roots of our country, we will also lose the memories, stories, and history associated with the food they produce.[15]

Like Kent Whealy, William Woys Weaver, author of *Heirloom Vegetable Gardening,* is an advocate of seed saving who strongly believes the health of our food supply is inexorably connected to plant biodiversity. Weaver stumbled into seed saving when he inherited a large collection from his grandfather. At first he was only mildly aware of the significance of the gift. Over time he came to understand that he was very likely the only person on the planet in possession of certain plant varieties, at which point he thought to himself, "Oh my God, I have got the ark right here, I had better learn how to sail it!"

Since then Weaver has worked with the Seed Savers Exchange, supplied seeds for historic gardens across America and Europe, and bred a number of new varieties of fruits and vegetables, including a drought-resistant tomato that is currently being grown in South Africa. During our interview he told me, "Genetic diversity is nature's way of maintaining life as we know it. The genetic diversity that may be lost through the extinction of garden vegetables is an issue that spills over into the political arena, for it touches directly on mankind's ability to feed itself. The harsh lessons of the potato famine of the 1840s seem sadly forgotten."[16] The Irish potato famine creeps into virtually every conversation I have with seed savers and other agricultural conservationists. I wonder what makes industrial agriculturalists believe that its lessons are only relevant to our past and not our future.

We know that the potato blight rotted virtually the entire potato crop of Ireland during 1845 and 1846; this fungus eventually caused famine and disease and the deaths of over 1 million people. If the Irish population had not been subsisting on, effectively, one variety of one vegetable, many of the deaths might have been prevented. Today's lack of diversity and monocropping could result in a similar disaster, and in fact a permutation of the same fungus has caused significant damage to America's potato crop. Clearly we need to learn from our history, or it will become our future.

Sustainable Trash

One of the challenges for any type of sustainable system is waste management. Chemicals, agricultural runoff, manure, unused food products, and packaging materials present problems for every segment of the industry. Restaurants, farms, manufacturers, and producers all face skyrocketing disposal or cleanup costs, and virtually every disposal method available has a negative effect on the environment. The true challenge is to find a way to do the least damage. Some creative individuals and businesses have implemented unique waste management programs that provide excellent examples for anyone in the food industry.

The nonprofit Intervale Foundation, based in Burlington, Vermont, was organized in 1988 by Gardener's Supply, a small horticultural company started in 1983 by a handful of enthusiastic gardeners. Today it is the country's only gardening catalog with an in-house research and development team, community farm, and display garden. Located in a flood plain, the land that now belongs to the Intervale Foundation was originally part of a farm, long since abandoned due to land misuse, including overproduction of corn and pesticide pollution. When Intervale took possession of the land, it began by establishing the Compost Project, which accepted food waste from area institutions such as Fletcher and Allen Health Care and the University of Vermont, as well as local restaurants. Today, the seven- to ten-acre composting center has several full-time employees and turns a healthy profit for the foundation. In 1997 it received 10,000 tons of waste, previously destined for landfills from area farms and dairies, Ben & Jerry's, and home lawns. Five thousand tons of finished product were later sold to local gardeners, landscape artists, and nurseries. Profits helped subsidize Intervale's many agricultural projects.

One of those projects, Green City Farm, is the ultimate model of the cycle of sustainability. Food production at the organic farm begins with compost, some of which is derived from food waste produced by Fletcher and Allen Health Care, and ends with the sale of produce right back to the hospital. Foundation director Daphne McPherson told me that "in 1997, the hospital composted eighty tons of waste from its food preparation and steam tables, and this was transformed into forty tons of compost.... During the summer, it received in return ten tons of hours-old produce at a wholesale price of $6,000."[17] In addition to Green City Farm, Intervale is home to six other projects; one, the Intervale Community Farm, feeds over 300 families and is the largest CSA (community supported agriculture; see p. 60 for more on CSA) in Vermont.

One of the challenges that all restaurants face is garbage. This is an area that Annie Somerville, executive chef of Greens restaurant in San Francisco, feels strongly about. A vegetarian restaurant produces a tremendous amount of "wet" garbage, and at Greens (see p. 233) about 50 percent of it is composted at Green Gulch Farm. When the farm delivers the restaurant's produce, it also removes the wet garbage, which it uses later as compost. Annie believes that proper handling and disposal of refuse is an integral part of a sustainable food supply, and she has become passionate about her recycling efforts. In addition to sending food waste to Green Gulch, she also recycles cardboard, glass, cans, plastic, and paper. She told me, "The city supports our compost initiative. I asked the two fellows [city employees] that came out and talked to me, " 'Out of all these people who have accounts with the garbage company, how do you know who is really recycling?' He replied, 'In your case, you were really excited about garbage!' And I thought, 'That is great! I am excited about garbage!' "[18]

As manager of natural resources at Ben & Jerry's, Andrea Asch is responsible for the company's environmental operations. High-priority goals for her are the reduction of packing materials generated by internal production and facilitating the transition to chlorine-free paper packaging for all Ben & Jerry's products . According to Andrea, the research and development team is experimenting with large recyclable containers for their ingredients that, when emptied, could be returned to the supplier for refilling. As an example, she explained, "We buy all of the cookie dough pellets that go into cookie dough ice cream, our number one seller, from Rhino Foods here in South Burlington. For years we have been buying the cookie dough pellets in a plastic bag that is inside a cardboard box. They ship thousands of these to us a month. After we utilize the cookie dough pellets we dispose of the bag and the cardboard. We are working with them now to put the pellets in a reusable larger tote that could be easily shipped. They would come in, we could dump them into the ice cream, and send the tote back for washing and reuse."[19]

In addition to cutting nonrecyclable products within production facilities, Ben & Jerry's is beginning to use unbleached paper products to reduce dioxins in the environment. Dioxins, some of the most carcinogenic products known to man, are compounds created during the paper-whitening process. Ben & Jerry's strongly believes that there are safe and practical alternatives to chlorine bleach: in their case, simply to use unbleached paperboard. Although not yet fully implemented, both programs are in the prototype phase.

Transportation and Fossil Fuels

Transportation of goods is one of the most difficult issues for me to come to terms with in the discussion of sustainability. I am a firm believer in buying locally, regionally, and seasonally, but I am also the corporate chef of the Putney Inn, a 350-seat facility serving three meals a day, 365 days a year—in Vermont. My primary responsibilities are to satisfy the needs of our guests and at the same time make the restaurant profitable. While my menus are generally seasonal in nature, I often have to purchase food from as far away as California. As I considered the sustainability of shipping foods cross-country (or from one continent to another, as some chefs often do), I wondered what the hidden costs are—emissions, use of fossil fuels, and loss of income to the local community.

In *Rainforest in Your Kitchen* Martin Teitel addresses the environmental costs of food transportation: "According to one researcher, it takes an energy expenditure of 435 calories to fly a five-calorie strawberry from California to New York."[20] I am at once horrified and caught firmly on the horns of a dilemma. What am I to do when the restaurant owner wants fresh strawberries in winter because our guests want fresh strawberries in winter? What is the ultimate cost to my restaurant and community if I refuse to serve certain items at certain times of the year and lose a portion of my clientele? Having regional, in-season foods is no good if no one comes to the restaurant to eat them, and educating the customer is a long slow process.

Here again, an issue that seems clear at the outset (do not buy fresh strawberries in winter, for example) turns out to be complex and difficult to resolve. Our culture has evolved in such a way that when dealing with matters of sustainability, one is required to perform a delicate balancing act, measured not in the black and white "good" and "bad," but in grayer "better" and "worse."

George Southworth, CEO and general manager of Northeast Cooperatives, a full-service wholesaler supplying more than 10,000 natural products to over 1,400 accounts in New England, has given some serious thought to this very issue. When I asked him how he reconciled the use of fossil fuels to transport foods with sustainability, he responded, "We look at sustainability in a lot of different ways. One is the sustainable system. [It needs to be] financially healthy to be sustainable. Another way to define [sustainability] is low input of energy resources. [In an effort to balance the two] we just turned over our trucking fleet . . . again, and it was only three years old. Trucks with advances in computer systems that monitor and drive the engines as well as improvements in refrigeration use less energy than three years ago. So we have actually made this conscious decision [to turn over our fleet] every so often."[21]

George explained that for him there are a number of complicated issues. Glass is a good example. Everyone assumes that glass is a better alternative to plastic for bottling milk and juice because it is more recyclable. Unfortunately, the cost of shipping heavy glass bottles from the East to the West Coast and then back again is much greater than the cost of shipping lighter, polymerized substances. They may not be as efficiently recyclable as glass, but they are recyclable *and* it costs less for the business as well as the environment. Of course, ideally we would not ship goods from coast to coast, and we should reevaluate this practice. However, in reality we do it all the time, and it is better for companies like Northeast Cooperatives to make thoughtful choices like this one than to ignore the issues entirely.

Farm and Kitchen Workers

We have touched on how the factors surrounding sustainability affect businesses, but have not factored in those who work to make businesses run. On this subject Michael Ableman's views are instructive:

> Even at its best, farming is extractive. It consumes resources, both natural and human. Sustainable agriculture is often discussed in terms of the soils, air, and water. It rightfully addresses the distance food must travel and the impacts of farming on the environment. We must also look at how well it sustains the people who do the work. It is a struggle to provide good wages, quality housing, health benefits, and a sense of ownership from a business that earns its annual budget by the pound.[22]

The link between community and sustainability is firmly established in the minds of many with whom I spoke. Bob Anderson, president of Walnut Acres Organic Farms, chairman of the National Organic Standards Board (NOSB), and former president of the Organic Trade Association, is one such person. In 1946 husband and wife Paul and Betty Keene quit teaching and borrowed the money to purchase Walnut Acres, one of the first organic operations in the United States. Their inspiration had come during a trip to India, through the teachings of British scientist Sir Albert Howard, who had become known as the "Father of Organic Farming." The Andersons, Bob and wife Ruth, inherited the business from Ruth's parents and have kept the farm chemical-free for over fifty years. Bob's hope is that industry professionals can help the consumer understand sustainability in a broad sense—that is, understand that the economic viability of the farm and the farmer is of the utmost importance: "I think that if we can, in sustainability, realize that we are also enhancing the fabric of rural communities and creating opportunities for small family farmers, then we have really accomplished something."[23]

Livable wages and safe work environments are imperative for farmers if an agricultural system is to be considered sustainable. Unfortunately, U.S. farmworkers' living standards are among the poorest in the country. The very people who cultivate food crops for the rest of the nation, even the world, frequently go to bed hungry, sleep under leaky roofs, cannot afford to heat their homes, and have no access to decent health care. A 1997 USDA Economic Research Service report by Jack Runyon states that the median weekly earnings for hired farmworkers in 1994 was $238. Many are living at or below the poverty line.[24] Additionally, farmworkers may have been routinely exposed to carcinogenic pesticides and herbicides without even knowing it. In *Food for the Future*, Patricia Allen wrote that prior to 1992, pesticides were "exempt" from hazardous information disclosure. Particularly at risk were children, who often had to "work in the fields to earn income to contribute to family survival."

At that time, a New York study disclosed that "40 percent of farm worker children studied had worked in fields still wet with pesticides."[25] The Worker Protection Standard (WPS), issued by the U.S. Environmental Protection Agency in 1992, now requires employers to take steps to reduce the risk of pesticide-related illness and injury by supplying information about exposure to pesticides used in the production of agricultural plants on farms and in forests, nurseries, and greenhouses. Under the law, employers must explain how pesticide exposure occurs and describe ways to avoid and/or lessen exposure. Unfortunately, children could still be at risk. EPA pesticide tolerances are set for adults with no provisions yet being made for children.

Kitchen workers also frequently labor under terrible conditions. American culinary tradition has its roots in European kitchens, where young men were placed in apprenticeships that required them to work from dawn to dusk, in some cases even sleeping in a corner by the stove until their next shift. Today's kitchens are certainly less grueling, but the traditions of long hours, brutal physical labor, and the requisite dedication of one's life to the kitchen god today remain as holdovers from the past. Also, like farmers, cooks are rarely given full health benefits, salaries are incredibly low (most kitchen workers still punch a clock and do not make much more than $8 to $10 per hour on the high side), and paid time off—vacation, maternity leave, sick days, or personal days—is virtually unheard of. Regrettably, this unsustainable way of working has pushed many talented culinarians out of the industry.

Rick Bayless, chef/owner of Frontera Grill and Topolobampo in Chicago, feels strongly about creating sustainable conditions for workers in the food industry. He finds it ridiculous that kitchen workers "have to be able to endure a medieval apprenticeship, work eighteen hours a day, and have no outside life at all. . . . In our restaurant, we try very hard to make sure that we have a sustainable restaurant from beginning to end. We pay people living wages, and do not require them to work . . . more than fifty hours, and most people clock in between forty and forty-five. We have two days off a week because the restaurant's closed two days a week."[26]

Eve Felder, chef-instructor at the Culinary Institute of America, agrees with Bayless. She believes that being sustainable means taking care of one's community, customers, and employees—equally. Felder feels that the traditional kitchen environment is abominable. In our interview she told me, "You work outrageous hours for very little pay, for no health benefits, for no retirement, and very rarely vacation. I think that we need to realize that if we train incredible cooks, and we have a very strong team that understands the food, they know the farmer, they are involved in the restaurant, we are going to serve much, much better food. And it is going to be done with energy, and love, and our customers will feel that."[27]

Like Felder, I have been in the food service industry for a number of years, and as a matter of personal experience, I can state, unequivocally, that the traditional blatant disregard for the health and well-being of kitchen workers must change. To date, we have made infinitesimal progress in this area. Feeding people is all about giving, sharing, and nurturing. How can we possibly expect that from kitchen workers and cooks who continue to be exploited?

SUSTAINABLE FARMING PRACTICES

Sustainable farming practices come in many forms—certified organic farming, integrated pest management (IPM), and biodynamic/biointensive farming, to name a few. Each is unique in the details of its approach to sustainability, but all are low-input (chemically) and utilize pest and weed management systems with little or no chemical additives. Most rely heavily on local and regional markets to generate sales and espouse a continued focus on replenishing the soil and maintaining natural resources.

Walnut Acres Organic Farms is heavily involved in keeping its business local and sustainable. Bob Anderson explained to me that it all began in the farm's infancy:

> We were raising small grains and as there was no organic marketplace for them in the early 1950s, we had more than we could use—so we started milling flour. As we milled flour, the logical extension of milling flour was to start making baked products, like cookies, breads, and granolas. The more milled products that we made, the more milled by-products or waste products that we had, and we realized that the logical extension of that then was to start raising livestock because we could feed them the mill cleanings. And so we now have this farm that raises about 100 head of steer, that produce the manure, that close the loop again, and make this a very sustainable entity unto itself.[28]

Walnut Acres is a wonderful example of a farm that combines a sustainable system of caring for the land and providing for the future needs of the community and its resources. As its business has grown, so too have the facets of its operation, which have also provided more jobs for the community, a secure rural lifestyle for its employees, and high-quality, clean food for people across the nation.

The shipping of products all over the country and the use of fossil fuels, as we mentioned earlier, are issues directly related to sustainability. Walnut Acres has spent over fifty years providing organic food to its customers, and in the spirit of sustainability, the Andersons have tried hard to balance all aspects of their business—including the use of fossil fuels. Without shipping, many people would not have access to the variety of organic products offered by Walnut Acres, so Bob has arranged to have the United Parcel Service pick up large shipments with semi-trucks, rather than carry several smaller loads a week. There is no way around the cost of shipping, but his choice has proved to be a good compromise.

Cheese is a mainstay of Vermont agriculture, and not two miles from the Putney Inn, where I work, is Major Farm, a small sustainable producer making some of the best sheep's milk cheese in the country. In fact, their Vermont Shepherd's Cheese was honored as Cheese of the Year by the American Cheese Society. Cindy and David Major patterned their farm after the small cheese producers they saw in Europe. They are currently milking 160 ewes, which yield about 9,000 pounds of cheese. In 1998 they completed construction of a cheese-ripening cave and started a regional coop of sheep dairy farmers known as Vermont Shepherds. In an effort to farm sustainably, members of the coop have agreed to milk their sheep only in the spring, summer, and fall. Cindy told me that grass-fed sheep produce a much more flavorful cheese. Additionally, she pointed out that the winter hiatus for the sheep benefits business operations. As she said, "It really works out well for seasonality; we have six months of the year when we are busy producing milk and cheese, really concentrating on production, then the other six months we are focused on marketing. Before we know it, it is time to get ready for the next season."[29]

Another area cheese producer, Vermont Butter and Cheese (VBC), began in 1984 as a dream for Allison Hooper and her partner, Bob Reese, after Allison completed a cheese-making apprenticeship in Brittany, France. Today, what started with a small herd of goats has grown to be an exceptionally successful and sustainable business. As their fledgling business took its first steps in the mid-1980s, it did not take long for Allison and Bob to realize that their little herd of goats would not be able to provide them with a decent living. As they pondered their options, realizing that they would have to team up with other area goat farmers, there was a dramatic turn of events in the cow dairy industry.

Traditionally, dairy farmers transported their milk from farm to creamery in ten-gallon stainless steel cans, but a sudden shift away from milk cans resulted when tanker trucks became the preferred method of transport. Larger trucks were having a more difficult time reaching more rural farmers, and even if the trucks could get to them, many of the small hill dairy farms could not afford to convert their operations to meet the new market requirements. This led to industry consolidation as well as the migration of the hill farmers to the bottomlands. Many hill farms were left vacant, their small-scale equipment unused. Allison told me that the remaining "wonderful little plots of real estate around Vermont, where empty barns held little milk houses and very small bulk tanks," were perfect for goat farmers. The small cow-farming industry gave way to the goat dairy industry, and VBC has played an important role in its continued success.

Most of the goat's milk used by Vermont Butter and Cheese comes from eighteen local farms, each of which has at least 100 goats (the number most farms require to be financially viable). Over the years, VBC has cultivated a completely symbiotic relationship with the farmers. As an example, some cheese makers only buy milk as they need it. One week their farmers may get paid for 10,000 pounds of milk, and then for three in row they may see no income at all. It is better for the cheese maker, but devastating to the farmer. Vermont Butter and Cheese takes all the milk its farmers produce, whether it is needed or not.

The company's demonstration of its level of commitment does not end there; Allison told me that they "spend a lot of time with the farmers, sort of walking them through the management issues they need to address in order to survive." They have realized that if they want to be in partnership with the goat farmers, it has to be a true partnership in every sense of the word. Over the last fourteen years, Allison and Bob have created a successful and sustainable business for both their farmers and themselves. In 1996 VBC received recognition by the U.S. Small Business Administration as Vermont Small Business of the Year.[30]

Dairying, as a concept, seems idyllic to me. *Bucolic* and *pastoral* are the words that spring to mind as I ride my bike along Vermont's back roads, past loosely fenced, verdant, cow-dotted hillsides. *Organic* seems like an appropriate descriptor for milk products, but I have learned that in conventional dairy farming, nothing could be farther from the truth.

Peter Flint, now coowner of the Organic Cow, began his career in 1972 as a conventional dairy farmer in Turnbridge, Vermont. By 1989 Peter and his wife, Bunny, had grown so disturbed by conventional dairying practices that they decided to go organic. Bunny told me that "most conventional dairy herds [are] fed high-energy feeds containing dried [beef] blood, tallow [also a beef product], cottonseed meal that has been heavily sprayed with pesticides and insecticides, as well as a host of antibiotics." Most cows are also given hormones to stimulate "artificial" heat, which, against the cow's natural cycle, forces the production of multiple eggs in order to allow for artificial insemination. Bunny believes that "there is a lot of hormone work done every day on a conventional farm that people are not aware of, and all of those hormones are going into the milk."

Bunny and Peter faced numerous challenges in transitioning the herd from conventional to organic dairy farming. One initial surprise was the cost of organic feed: it costs $100 to $150 more per ton to provide the Flints' cows with feed that has no beef or animal by-products or chemicals. But the initial trouble they had paying for organic feed was well worth it, Bunny says: "When we transitioned the herd to organic, one herdsman said to us, 'We have to wake these cows up to milk them, they are just so mellow.' We just noticed this incredible difference in our animals, not just their well-being and general health, but their vitality as well; we never really knew our cows until we knew them organically."[31] The Organic Cow grows its own organic hay, uses rotational pasture management practices, and allows its cows to forage rather than feeding them in a feedlot. The end result is a chemical-, hormone-, and additive-free product that truly tastes great. One hundred years ago this kind of milk was practically standard fare. Today, it is an anomaly.

Another type of agriculture that is bringing back ancient practices is biodynamic/biointensive farming. Introduced to Western culture by Rudolf Steiner in 1922, it is largely based on a 4,000-year-old Chinese method. Philosophically, biodynamic/biointensive farming is based on the idea of "working with the energies which create and maintain life"[32] and is patterned after our land's natural regenerative structure, in which multiple species grow in close proximity while protecting and fertilizing each other. This form of agriculture strives to produce the optimum yield in the smallest amount of space by forgoing walkways and using a system of raised beds.

In the United States, John Jeavons, author of *How to Grow More Vegetables on Less Land Than You Can Imagine*, has been a strong proponent of this system. He founded Ecology Action in 1970 as an environmental research and education "organization which has revolutionized mini-farming around the world." It has become a vehicle for promoting the advancement of biodynamic/biointensive farming as a way of building soil, reducing water and fertilizer consumption, and increasing yields.

Twenty-five years into its research, Ecology Action maintains that in practice, composting and synergistic crop combinations can build up soil sixty times faster than in nature, while making possible a 67 to 88 percent reduction in water consumption and a 50 to 100 percent reduction in purchased nitrogen fertilizer. Its studies also seem to indicate that biointensive farming can produce an average of two to six times more vegetables per acre, with an energy consumption of 1 percent of that used by conventional farming. Jeavons's research may prove particularly important in the years to come—especially if urban growth continues to swallow up what is left of our nation's farmland.

Another major challenge to farming, and sustainable agriculture in particular, is seasonal climatic variation. The growing season in the Northeast, for example, is significantly shorter than it is in California, Florida, and even much of the Southwest, but restaurant guests and grocery shoppers still like to have fresh green salads in the winter. As we already discussed, most restaurateurs (and grocery stores, for that matter) ship their produce cross-country in winter, but one northeastern farmer, Eliot Coleman, is giving New Englanders a choice. Not only that, everything he produces is organic.

Coleman farms from October to May in over 12,000 square feet of covered greenhouse. When I spoke with him, he explained to me that as the temperature drops in the late fall, he drags a portable greenhouse over his produce, and as the season progresses, another layer of insulation is added inside the greenhouse. The effect, he said, "is the same as moving one and a half USDA zones to the south. . . . So I [start out] here in Maine, or Zone 5. When I walk into my plastic greenhouse . . . I am [effectively] 500 miles farther south, in New Jersey, as far as winter climate goes. If [I] then put a second layer inside that plastic tunnel, about a foot above the ground, that area [is] another 500 miles to the south, or the equivalent of being in Georgia. And this is for free as regards fossil fuel consumption and heating costs. It is that simple. That is what has been so utterly delightful about this." When I spoke with Coleman, he told me that he was inspired to pursue this type of agriculture as a result of a trip he and his wife, Barbara, took to Europe. He explained,

> Just before we started doing this commercially, we took one last research trip, and followed the 44th parallel of latitude, across France and Italy. And the reason we chose the 44th parallel is that's the parallel of latitude that we live on right here in Harborside, Maine. People assumed we were going to end up in Oslo and go skiing, but Europe is so much further north on the globe than the United States that the same latitude we live on here runs along the Mediterranean coast of France, and Italy, and on through Tuscany. We have as much winter sun here in Harborside, Maine, as they do in Avignon. Portland, Maine, has the exact same amount of winter sun as they do in Cannes!

Coleman told me that understanding where his farm was geographically helped him to understand what would grow there. He and his wife sell lettuces and winter greens (like watercress), carrots, melons, tomatoes, peppers, eggplant, and European cucumbers, and they make a profit. Eliot said they "should be able to expect a growth of $5 a square foot." Over the course of the year, from the blend of crops they grow, on 10,000 square feet of land they should bring in close to $50,000. He continued, "If we had more than $15,000 worth of expenses we would be idiots, so that means that the two of us, and it sure takes two of us to run this thing, end up with $35,000 for that eight months' work." Not bad for a small vegetable farmer in New England, and especially interesting for one who takes summers off.[33]

Integrated Pest Management

Integrated pest management is a system of insect control often implemented as part of a sustainable agricultural system. The management system itself is not considered organic, but it does promote minimized pesticide use and enhanced environmental stewardship. On a basic level, IPM strategies include careful timing and specified use of pesticides. Apple farmers in the Northeast are especially interested in IPM, because while many of them might like to grow organic apples, it is difficult in that region, where a disease called "apple scab" is prevalent. With IPM, farmers monitor pest activities by placing traps in the orchards and monitoring, identifying, and recording the number of insects in residence. Rather than spraying a "blanket" of chemicals to kill a wide variety of insects, IPM encourages the application of insecticides that target specific pests in residence. It is not an ideal system, but it is certainly less harmful to consumers and the environment than conventional methods.

Community Supported Agriculture

CSAs are literally farms supported by their communities. Small farmers are increasingly looking to community supported agriculture (CSA) to support them in their efforts to continue farming in a way that is sustainable to the land, their families, and the community at large. Currently there are about 650 CSAs around the country, most situated on urban outskirts. Participants are ordinarily asked to choose among a variety of membership options. A person, or family, is required to purchase a share before the season begins, which entitles the purchaser to weekly or biweekly pickups (in some cases, deliveries) of fresh, often organic, produce.

The average cost per share varies according to region and the consumer's level of participation on the farm during the season, but most family shares can be purchased outright for an average of about $460. If a person is willing to work on the farm a few hours a week, he or she can purchase a part-time share at a reduced rate. A full-time share requires the purchasers to work many more hours but entitles them in most cases to an entire season of free produce. No matter which option a person chooses, he or she is getting an incredible deal. The USDA's Sustainable Agriculture Research and Education program (SARE) has determined that the average CSA share yields approximately 400 pounds of produce for the season. Consumers buying an equivalent amount of organic produce in a store could expect to pay upwards of $1,000. The same amount of conventionally raised product would cost between $680 and $780, translating to a $200 to $450 savings for CSA members, based on the above-mentioned average share price. Setting the system up this way is clearly beneficial to the consumer, who has a new opportunity to become part of an agricultural community through hands-on farming experience, special classes for kids, harvest days for the entire CSA, and even periodic picnics or other organized activities. Most CSAs even distribute a weekly or monthly newsletter reporting farm activities and various other events. Most importantly, however, it helps the farmers. Selling shares in advance gives farmers the necessary capital for purchasing seeds and supplies. It also saves biologically diverse farms that might ultimately disappear due to an inability to compete with large monocropping farms and the temptation to sell out to agribusiness.

CSAs are not limited to rural areas. In New York City, consumers looking for organic food products can utilize the services of Urban Organics, a distributor of organic produce, groceries, and dairy goods. Organized more along the lines of an organic food club, Urban Organics asks its customers to purchase a membership, which allows them to choose from a menu of items. Boxes of fruits and vegetables are provided in various sizes to suit the needs of a particular household. As with typical rural CSAs, what arrives at the consumer's door is dependent upon the seasons. Unfortunately, however, members do not have the option to work on the farm.

Some CSAs also participate in low-income farm shares, which are subsidized by various organizations around the country. The Northeast Organic Farming Association (NOFA), directed by Enid Wannacott and dedicated to "local diversified organic agriculture," has been in operation since 1970. A nonprofit organization with a current membership of 700 working out of seven independent state offices, NOFA sponsors a wide variety of projects, including certification of organic products, an agricultural education program geared to integrating curriculum into elementary schools, and a program that supports organic produce in school lunch programs. Additionally, NOFA oversees a Farm Share program, which supplies low-income families with CSA shares, giving at-risk households the opportunity to eat fresh, regional, organic produce. In Vermont, the annual Share our Strength (SOS) Benefit funds the Farm Share program. Wannacott would like the program to go even further: "What we are really working on is, how do you make that program more of a permanent solution and less of a Band-Aid? We are working with the families to develop gardens, working with those families to be mentors for the next set of families that comes in. We have also overlapped with Operation Frontline, to do classes in nutrition education, and how to keep purchases local." This again epitomizes a sustainable community-oriented program.[34]

SUSTAINABLE BUSINESS PRACTICES

It is difficult, if not impossible, to find a company with a more socially conscious agenda than Ben & Jerry's. Since Ben Cohen and Jerry Greenfield started the ice cream company in 1978, Ben & Jerry's has been inspiring humanitarian activism throughout the United States, but it was not until

1988 that they began including a "Social Report" as part of their annual report. In 1992 the company signed the Ceres principles, which state that they will "operate all aspects of the business as responsible stewards of the environment by operating in a manner that protects the earth."[35]

In 1997 Ben & Jerry's adopted a social mission statement entitled "Leading with Progressive Values across Our Business," in which the company's position on sustainability is made clear: "The growing of food is increasingly reliant on the use of toxic chemicals. We support socially and environmentally sustainable methods of food production and family farming."[36] Ben & Jerry's appears to be holding true to its goals, a major reason why many of the company's employees decided to join the Ben & Jerry's team. Andrea Asch and purchasing manager Todd Kane both affirmed that the company's social policy was a primary consideration in their applications for employment. Kane told me that when he was looking at Ben & Jerry's as a potential employer, he was strongly influenced by the fact that "it is a company that has a value system with which I am aligned—I can have the same set of values 24 hours a day, not one set at home and a different set at work."[37]

I spoke with Todd and Andrea about how Ben & Jerry's supports sustainable food supplies and chemical-free ingredients. One of the most important relationships the company has is with St. Albans Cooperative Creamery. All the dairy products for Ben & Jerry's ice cream come from the coop owned by Vermont family farmers. The company pays a premium for the unwavering guarantee that every drop of the milk and cream it purchases is rBGH-free.

Ben & Jerry's feels so strongly about the public's right to know they are buying rBGH-free ice cream that it recently went to court to secure the right to label its ice cream as such. Winning this lawsuit, brought against the state of Illinois, paved the way for Ben & Jerry's to inform its customers about its policy on rBGH. As Todd explained, the company is equally careful about the rest of its ingredients. "When sourcing new ingredients, I work in conjunction with our quality and product development group. The first thing we tell potential suppliers is that our product will be labeled all-natural. We do not allow any synthetic ingredients, artificial colors, or artificial flavors; we simply will not use them. It is best not to submit samples that contain them."[38]

Although aside from dairy products, Vermont ingredients are not the majority of purchases for Ben & Jerry's, the company continues to have a positive effect on the state. From packaging to philanthropy and politics, I cannot think of a company that has a more beneficial impact on the state, its immediate local community, and the people of Vermont.

George Southworth at Northeast Cooperatives is also doing his best to contribute to the sustainability of local businesses by helping to expand markets for regional products. He told me that Northeast Cooperatives has been a driving force behind the success of the Organic Cow. Southworth had seen a steady increase in the demand for organic milk, and the coop was bringing in their organic milk from the Midwest. George knew the Flints through their connection with Northeast Cooperatives prior to their establishing the Organic Cow as a commercial organic dairy. He wanted to help them grow their business and find a supplier of local organic milk for his clients. He told Peter and Bunny Flint that he believed they could work together to create a fluid milk program in which they could use their milk cartons as "billboards" to promote the uniqueness of the Organic Cow and to market their other products. He explained his involvement, in our interview:

> We helped them create the business plan, we gave them a guaranteed minimum amount we would purchase, we gave them a guaranteed price—we set them up with fiberboard suppliers, the milk carton—directed them toward people who could supply the machinery. We took a 6 percent margin, which is lower than the margin that we normally take on a product, and I also did a wild thing, which was put it on allocation. So we have standing orders with all the stores, which is a lot to manage, but they are guaranteed supplies to the different stores over a period of time, and we went out and sold it for them.[39]

The way in which Northeast Cooperatives helped promote a local business is the very essence and spirit of sustainability. In this case, the Flints' business grew in extraordinary ways in part because George Southworth and Northeast Cooperatives took the time and initiative to expand the consumer base for an otherwise completely local enterprise. If more people operated their businesses this way, smaller communities would have a greater chance of sustaining themselves in the marketplace. However, even the best intentions often go awry, as businesses grow ever larger. The Organic Cow was eventually bought by Hood Dairy, and more recently by Horizon Dairy, and while still producing a quality organic product, it is much less of a small community icon.

THE CHEF-FARMER CONNECTION

It is impossible to discuss the chef-farmer connection without first introducing Alice Waters, one of the strongest advocates of regional and seasonal restaurant food in America, who virtually pioneered the current relationships between chefs and their suppliers. Many of today's culinarians consider her restaurant, Chez Panisse, the mecca of organic California cuisine, and it has ultimately proven to be the training ground for many of this country's great chefs. Its beginnings, however, were much more humble than its current reputation. As a young woman Waters journeyed to France, where she fell in love with the local markets, the fresh produce, and the farmers who made it all possible. Upon her return to California, she began showing off her newly acquired cooking skills by throwing dinner parties for her friends. Twenty-five years ago, what began as a hobby metamorphosed into Chez Panisse, where Alice insists on utilizing only the most exquisite regional and seasonal ingredients. From the beginning, the restaurant employed a forager whose mission it was to find the freshest local, seasonal, and, whenever possible, organic ingredients. That was not an easy task over a quarter of a century ago. Alice has worked hard through the years to cultivate lasting friendships with the farmers who supply her restaurant, believing that "good food starts in fields and orchards well tended. This is knowledge that we ignore at our peril, for without good farming there can be no good food; and without good food there can be no good life."[40] Alice's was the first politically motivated voice heard above the fast-paced din and clatter of the culinary world.

Waters's voice still cries out to the world to slow down and consider the choices we make when we purchase our food. She believes that there is much more to satiating a grumbling stomach than simply ingesting food. For her, eating is a political act. As she explains it,

> The decisions you make are a choice of values that reflect your life in every way. Buying Big Macs from the people McDonald's buys its meat from, who are raising these cattle or kangaroos or whatever goes into what they call beef, is the complete opposite of the way it should be done. When I buy food from a farmer, I know who he is, I know he cares about my well-being, and I know he has taken care of the land he is farming. I have a responsibility to him and he to me. I could not put the food I cook on the table without him, so I really treasure this relationship.[41]

It comes as no surprise, in light of Alice's unyielding political activism, that the James Beard Foundation honored her as Humanitarian of the Year in 1997 for "advocating local, sustainable agriculture and championing the philosophy that our sense of community should come from an understanding of and respect for the earth."[42] As do many like-minded culinarians, I feel her contributions to the culinary world have been tremendously beneficial to chefs and cooks around the country.

After traveling from coast to coast working with some of the country's most renowned chefs, including Alice Waters and Larry Forgione, Melissa Kelly was the executive chef at the Old Chatham Shepherding Company Inn in upstate New York until 1999. Many of the chefs with whom she trained had unique working relationships with their suppliers, and Melissa learned early in her career to place a premium on quality ingredients. At the Old Chatham Shepherding Company she could hardly have been closer to the people who produced her food; a large percentage of her ingredients came directly from the inn's farm and creamery. The inn, which features luxury guest rooms, the restaurant, a farm, and a creamery, was founded in 1994 by Nancy and Tom Clark. Five short years later, with a herd of over 1,000 ewes, they are the largest sheep dairy farm in the country. The 500-acre farm produces milk, cheese, yogurt, and meat, as well as a variety of fresh greens, vegetables, and herbs specifically for the restaurant.

As a visitor to the farm and restaurant, I had a sense of the unqualified connectedness among the animals, the land, the restaurant, and the food. Kelly's menu, which changed daily according to the availability of ingredients, was a testament to her tirelessly innovative and creative spirit. And, as expected, sheep's milk was ubiquitous, used in yogurt, ice cream, panna cotta, soup, and cheese. The farm's lamb remains a continual favorite, and in season almost all the produce comes from the inn's garden, picked, cleaned, cooked, and served by the kitchen staff. Before she left the inn, Melissa shared this story about how they were growing lettuce on the farm: "We figured out that what we needed to do in the greenhouse is have five boxes [because] it takes four weeks for the lettuce to get to the [perfect] stage. We do a mix [of lettuce varieties] in each box and we pick one box each week. By the time we get to the last box, the first one has regenerated itself and we can start again." Kelly relishes the control she has over the product: "It is really the ultimate way to cook, and it gives you so much flexibility. . . . You can pick something when it is perfect, or you can pick in stages. We will pick certain beans when they are really tiny, and then we will let them go to shell beans, and pick them again, so we have total control. I can say, 'I want the arugula just a tiny bit bigger,' and we can wait a day and pick it when it is just perfect."[43]

Melissa's experience as a chef who farms is virtually unique in the restaurant business, but, it is possible for chefs to cultivate similar relationships with their suppliers. Odessa Piper, chef/owner of L'Etoile, located in Madison, Wisconsin, has the dedication it takes to make farmers an extension of her restaurant. Odessa exemplifies the chef-farmer connection. Her passion for truly regional, seasonal, and organic food is contagious, and her menus epitomize the use of seasonal ingredients. I spent a couple of days at the restaurant with Odessa and her former sous-chef/forager, Tammy Lax, and saw firsthand the kind of resolve and tenacity a chef must have to be true to her priorities. Odessa told me that in 1983 she realized the importance of the chef-farmer connection:

> I understood that when my farmers had a strawberry crop, and there was a heavy rain, and a lot of sand was driven into the strawberries, the farmers were going to either lose the crop, or they would call me for help. I would just stop whatever we were doing and it would become the priority of the

day. We would figure out a way [and] we would be processing strawberries until 3 a.m., because it never was a question in my mind that the farmers' success was as important as mine. I could wash and process those strawberries, put them on my menu, and recover enough money to pay the farmer. I realized this [willingness for symbiosis] had become my philosophy.

Later that day, I had the good fortune to spend a morning walking through the market with Tammy as she searched for the perfect ingredients. It was an amazing experience. Tammy made her purchases and placed them carefully into little wood-sided red wagons, the backs of which were proudly emblazoned "L'Etoile." The food was unbelievably fresh and mouthwatering; as a chef it was inspiring to see, feel, and taste what the farmers had brought to market. The relationship that Tammy had with the farmers was one of mutual admiration and respect. At one booth I watched Tammy negotiate with one of the owner's children. The price the youngster quoted, Tammy thought, was too low, so she insisted on paying more. And the food was truly incredible, from sweet baby cherry tomatoes to melons and berries, carrots and beets, all destined to grace the plates of eager restaurant diners that very evening. When we returned to the kitchen later that day, the restaurant was processing tomatoes at the peak of ripeness for use in the winter. Tammy explained,

> We are processing those heirloom tomatoes that you saw [at the market] today. The good ones we are using right now for gazpacho, but the rest I roast with garlic, oil, and herbs, pull the skins off, and seed and freeze them. We are also processing Sungold tomatoes. We just slice those in half and dry them. We get 30 pounds at a time and at night when I am expediting on the line, I will be doing those between my other work. Each staff member on the line will also have a bucket, and they will be slicing between stuff that they are doing, and at the end of the night, we take them all, and put them into the dryer.[44]

Without this sense of shared responsibility and teamwork, Odessa's vision for L'Etoile would be impossible to realize.

During the summer months almost all of L'Etoile's produce is purchased through the farmers' market; everything is incredibly fresh and comes from area farms and producers. The most amazing display of Odessa's passion, however, is her dedication to using regional ingredients in winter—in Odessa's vocabulary, "being regionally reliant." In Wisconsin! She told me that at one point "local vegetable farmers were experimenting with wintered-over crops held under unheated hoop houses and [were] cellaring loads of squash and root [vegetables], and we bought every thing they had. [At that time] locally grown leafy vegetables, roots, syrups, nuts, eggs, poultry, cheeses, fresh and preserved fruits made up over 60 percent of our winter menu. This last winter [1997], with the exception of leafy greens, about 80 percent of our vegetables were locally grown." Out of curiosity I asked Odessa how much of her produce was organic, and her answer surprised me. She said that in the winter, the percentage is higher because she orders most of her greens from California, and organic is her assurance of quality. However, in the summer all of her suppliers are local farmers and friends, and she feels it is "more important to me [to buy from] small-scale [farmers], in season, [using] native seeds on native soils, or happily adopted species on native soils." Odessa believes that it is ultimately far more sustainable to buy quality ingredients from members of her community than it is to purchase solely organic product from people outside her community.

Animal products are no exception to Odessa's tireless pursuit of quality, healthful food. She shared this experience in trying to source veal for the restaurant:

> The story of L'Etoile trying to get humanely raised veal, fresh veal on the menu has taken years. Oddly enough, it all started with chickens. In order to get veal on the menu, we had to start with chickens. We still do not have a fresh supply of free-range poultry year-round, or even good, naturally raised poultry year-round, because, shock of shocks, this is one of those amazing splash-of-cold-water-in-your-face stories. I found out, I am a chef, a professional chef, who has been doing this for twenty-five years, and I just found out that chicken is a seasonal food! In the middle of winter, chickens do not like to put on weight, and they are cold, and they are not happy in the middle of winter, and it is not a good time to grow them.[45]

She started buying fresh free-range chickens from a local purveyor in the summer and during the winter purchased them from a national organic grower. In talking with the local chicken farmer, she came to realize that if she bought frozen chickens from him in the winter she could help support him through the slow season, which was important to her because she liked his product and wanted him to be around for next season. In return for her show of loyalty he agreed to begin growing "pink," or grass-fed (as opposed to milk-fed) veal for her. As a result of her tireless efforts, during her 1998 season Odessa was finally successful in purchasing locally grown lamb, bison, beef, poultry, and yes, pink veal.

Michael Romano is chef/partner of the Union Square Café in New York City. Manhattan is not the kind of place where one might think of finding a pillar of regional/seasonal cuisine, but some restaurants, like Union Square Café, are working very hard to be just that. In fact, it is largely through the efforts of Michael and owner/partner Danny Meyer that the farmers' market in Union Square is thriving. In 1985, when Meyer was searching for the perfect location for his restaurant, the Greenmarket's promise of "fantastic produce" was a major selling point.[46] During the years that followed, the market and restaurant developed a "cooperative" relationship that has served as inspiration for Romano's daily specials on many occasions. Romano says, " Our main focus is working with the farmers' market, at Union Square. We feel it is an important way [to] support local agriculture. We support our farmers [and] we support the principle of sustainability because we are using what is in season, locally grown." He and his staff walk through the market each of the four days a week that it is open. Just as Tammy does for L'Etoile, someone on Michael's staff goes to the market and physically hauls the food back to the restaurant. The food is frequently preordered, but just as often a special fruit or vegetable that catches the chef's eye that day might turn up on the menu as the daily special. "I was just walking into work the other day," Michael told me, "and talking to one of our farmers who had those beautiful pale green frying peppers, and I just had an urge to buy them, there is that spontaneous element to the market as well." Undoubtedly, there are challenges from a chef's standpoint in purchasing from a greenmarket, especially in New York City. I asked Michael what he felt were the most formidable aspects of his purchasing system. "I think the chef has to be willing to deal with vagaries, and changes," he told me. "The closer you get to the reality of nature, the more you have to deal with inconsistency, and real life-cycle stuff!" He believes that it is more important to accept and work with variances in nature than dominate the growing cycle with chemicals: "If you want 100,000 tons of broccoli all the time, you can get it, but there is a price that you are paying and that is not sustainable. The earth is suffering. We have to take care of it because it is what nurtures us."[47]

Ana Sortun, executive chef at Casablanca in Cambridge, Massachusetts, has been involved with local agriculture for a number of years. When she worked in the western part of the state, she would stop by the farms and pick up fresh produce on her way to work. At that time she learned the importance of developing good relationships with the farmers. Today she continues that effort through her involvement with Boston's "Fresh Sheet" project. I first heard Ana speak about the Fresh Sheet when she and I were on a panel together at a CC2000 regional conference at Allyson's Orchard in New Hampshire. The Fresh Sheet was originally conceived by Chris Schlesinger, chef/owner of the East Coast Grill, and Jodi Adams, chef/owner of Rialto, both in Boston. Under the auspices of pairing city chefs with regional farmers, the Fresh Sheet is a list of farmers, producers, products, availability dates, and in some cases delivery schedules that is regularly updated and faxed to participating chefs around Boston on a weekly basis. Ana's concept of sustainability stresses communication among chefs and farmers, who form relationships based on their common passion for food: "The farmers get inspired by what the chefs are doing with their food. They get inspired to have us come to the farms, and show us everything growing, and seeing our faces, it is a cool relationship. They come in here all the time and they go back in the kitchen, and they hang out, and they talk about the day's work, and see how we prepare their food, and that is really satisfying."[48] I have had similar experiences with purveyors, but the most memorable is with Kris Holstrum, an organic farmer in the mountains of Colorado. During a growing season that includes only about sixty frost-free days, she grows the most amazing produce 9,000 feet above sea level in the Colorado Rockies. For the past decade I have worked as the executive chef of the Telluride Film Festival in Colorado, during which we feed 15,000 to 20,000 people in multiple locations out of portable kitchens. For the past few years I have gone to Telluride in June and worked with Kris to decide what she would grow for the Labor Day weekend event. In late August, when I arrive to begin preparations for the festival, I confirm our original plans. The produce is harvested daily throughout the event, and when Kris makes her deliveries she closes the loop by hauling all of our "wet" garbage back to her farm for compost. As a chef, it is extremely satisfying to be involved in the process on that level.

Arguably the most well known vegetarian restaurant in America is Greens, located in San Francisco. Started in 1979 by the Green Gulch Zen Center, the restaurant has worked closely with the center's farm, Green Gulch, and its Tassajara Bakery since its inception. Renowned chef Deborah Madison ran the kitchen when the restaurant first opened its doors, and its current executive chef, Annie Somerville, trained with Madison before taking over her position in 1985. In *The Greens Cookbook*, Madison writes about the early days of the restaurant, "Working with Green Gulch made it possible for us to experiment with new varieties of food and to stay in touch with what is in season locally. We spent mornings on our knees at the farm pulling tomatoes, peppers, and eggplants—the farm hours added to an already long day of cooking, but the time spent was always generously repaid with the inspiration we gained."[49]

Annie and I picked a beautiful July day to tour Green Gulch. We met at Greens and drove to Marin, where the farm is located. I was instantly mesmerized by the wild plums and gardens of flowers and vegetables, magnificently sheltered by bamboo wind breaks. As Annie and I walked the farm that day, I got a firsthand look at how a sustainable relationship between a restaurant and farm works at its best. Greens buys as much of their produce from Green Gulch as they can supply, and just being able to go to the farm helps Annie write her menu. She elaborated, "I have not been hearing about the chard, I did not know that they had red kale, and I did not know that the leeks were still in the ground, so they will all crop up on the menu soon." In addition to their relationship to Green Gulch, Greens supports many other local farmers and producers. Annie told me that the majority of their products are locally produced and often specifically featured

Rhythms of Life for a Small-Scale Farmer

Kris Holstrum

What is it that farmers actually do? While we all deal with deadlines, setbacks, goals, and successes, the farmer's realm is a bit different. The weather, the seasons, the soil, and natural variables, like good bugs and bad bugs, most closely regulate our routines. What follows is a glimpse into a day in each season at Tomten Farm—a small organic farm and greenhouse at 9,000 feet in the southwestern Colorado Rockies. Farmers and growers all over the world face common challenges and awaken to similar daily regimens.

Winter begins the calendar year. In my region winter comes early and stays late, so I will pick a day in mid-January. It is snowing again. The kids have been bundled up and driven down the mountain to wait for the school bus. The morning chores are done. Paths of hard-packed snow wind along to the greenhouse and chicken coop. Wander off the path, and you would find yourself thigh-deep in powder. Thermometers in and outside the greenhouse have been checked and reset. The hens are happy, since their frozen water bucket has been emptied and replaced with fresh, cold liquid. It will freeze again soon, so they gather quickly and drink their fill. A few hens still lay eggs in the cold winter, but most will wait until it is a bit warmer before they hit the nest boxes.

In comes a load of firewood and off come the snow boots. A cup of hot tea helps take the chill off as chores yield to planning the next season's work. This is the time to review last year, reread journals, and incorporate notes into the new year's schedule. It is time to order seeds and supplies; growing areas must be mapped out, crop rotations figured, and visions nurtured. The more organized and thoughtful the plans, the smoother the implementation come spring.

The slightly calmer aspect of winter is one in which to read and reflect. The mountain of reading material has diminished since the snow covered the ground, but there is still a pile. If it is a harvest day, twice a week in winter, it is back out to the greenhouse once the risen sun has had a chance to do its magic. Row covers are pulled off the beds and an assortment of salad greens is picked and washed before being packaged for delivery. If my timing is right we will make our deliveries in time to pick up the youngest child from kindergarten. Some days the two of us will read in the school library until it is time to pick up the oldest, other days we will meet his bus at the bottom of the hill. We go through typical post-school snacks, homework, and activities before dinner. Winter days are short, and bedtime comes early.

Spring is a relative term in the high country. The equinox often brings big, heavy snows that defy our desire for warmth. A day in April will find us in the greenhouse again, after chores. This time we are planting. A short growing season means we need to have lots of transplants ready to go in the ground as soon as it emerges from under the snow. We will mix up a big batch of soilless mix using peat moss, vermiculite, perlite, and kelp. We will seed some new flats, then transplant tiny green plants, started two weeks ago, to larger containers. Each seedling is carefully teased away from the others, held gently by the leaves, and placed in the cells of the flat. The potting mix is firmed around the roots with good intentions, and when full, the entire flat is gently watered. The job is repeated, so eventually fifty kinds of lettuce, ten kinds of oriental greens, and lots of other unusual greens show their bright green faces.

By mid- to late May when the hardiest plants can go outside we should have ten to twelve thousand seedlings ready. Tending the seedlings and making sure things are ready to go is spring's most constant task. We also interview folks who have applied for our summer intern positions. In exchange for their labors, we give them a teepee and food, and we teach them organic farming and sustainable living techniques. While inviting folks into your life for the summer may not be for everyone, it works for us, and we have extended our "family" with marvelous folks.

Summer brings the rush. Up here we may get frost on the solstice, so tender crops are not planted until then. Planting, watering, and harvest schedules rule our lives. The kids are out of school, so there are no more bus runs. Occasionally they give us a hand planting and watering, feeding the chickens, and gathering eggs. Major harvests begin in late June, when we start supplying the grocery stores and a restaurant with greens and other homegrown goodies. A typical harvest day starts early. Breakfast at 6:30, then out to pick. At this altitude mornings are always brisk, but we have to get the tender greens picked and processed before the heat of the day. We will pick basil and tomatoes in the greenhouse. If it's not a harvest day we will spend an hour or two "in class" with the interns, giving them solid background in the topic of the day, be it soils, fertilizers, or sustainable philosophy. Then to work. We may be building a new compost pile or mulching beds with old straw to keep them moist. Every two weeks we spray the crops with kelp-and-fish-based fertilizer, and we always look for bugs or disease problems. We identify beneficial insects we see flying around and take some time to tend to our beehives. The farm sees lots of visitors, including local preschools and summer camps. These are fun but exhausting. We have the kids plant peas or potatoes and talk to them about where their food comes from. We also have to watch that they do not run rampant through newly planted beds or let the chickens out. The long days mean those endless tasks, like weeding, can go on past dinnertime. The desire to finish the job means I often have to be reminded that the family is hungry and "what's for dinner?" is my responsibility.

Autumn may be my favorite season. The long days of harvest, watering, and weeding begin to shorten. First frost here is usually mid-September but can come in August. School's back in session, so we have to roust the kids early as we all adjust to that routine again. The aspen leaves begin to turn gold, and we are treated to elk bugling across the valley when we are digging carrots or pulling radishes.

> ## Rhythms of Life for a Small-Scale Farmer (continued)
>
> While we look toward shutting down the outside production in October, we must plan and plant the big greenhouse for late fall and winter harvest. A typical day would find us pulling out frosted zucchini plants to throw on the compost pile. The piles are turned and prepped for winter. We will seed greens under the still-producing tomatoes in the greenhouse. It is time to gather the tools, make sure they are clean and sharp and not left in the fields. We eat well in the fall, and some of the most glorious days are spent in the kitchen, canning pickles or tomatoes—putting food by for the long winter days ahead.
>
> So this farmer is pretty much just like you. I worry about whether I am spending enough time with my kids and husband, and at the same time I wonder when I will ever get a chance to schedule some time for me. I hope the income meets the outflow. I organize the carpools and attend school events when I should be weeding or planting. But the rewards are rich. I live in a stunningly beautiful setting that I appreciate every day. My kids grow up eating fresh, organic food, and they know without a doubt where it came from and how much work it is to produce. They have the pleasure of gathering fresh eggs and the terror of being chased by the rooster. I do get a few private moments when I look up at the mountains surrounding me and take in a deep breath of thanks. Most of all, the delight and appreciation in folks' eyes when I deliver fresh, organic produce to them makes it all worthwhile. I would not be or do anything else.

on the menu. One of the producers, Strauss Dairy, has its milk featured prominently. "They do a fantastic job of producing organic milk, and we feature their milk on our list of beverages, and all our coffee drinks. You can buy a glass of Strauss Milk, and any of your coffee drinks will have Strauss Milk," Annie said.[50]

Chefs and farmers do not always have the time, or the inclination, to develop relationships like the ones mentioned above. Sometimes an external force is necessary in linking chefs and farmers. Roger Clapp, deputy commissioner of the Vermont Department of Agriculture since 1995, has been instrumental in creating a grassroots organization through his position as a representative of the government. He told me that he first became interested in linking chefs and farmers when he attended an agricultural workshop. He explains, "One of the presenters there showed how the percent of the consumer dollar that gets back to the farm had shrunk from 42 percent back at the turn of the century, to about 9 percent today. We looked at where that money was going, and the bulk of that percentage had shifted . . . from production to marketing. I realized that we needed to get farmers more involved with the marketing of their products, in order to recoup some of their lost market share."[51] The result was the Vermont Fresh Network, which in 1998 had a membership of 250. At the Putney Inn we include product and farm names on our menu and display pictures of the farmers in the restaurant. In turn, they distribute the inn's brochures at their farms. We jointly market each other's products in a way that is typical of most participating farms and restaurants. The system, funded in part by federal and state agencies, creates a direct link between farmers and chefs, who in turn pass information along to their customers in a way that would not be possible with traditional, mainstream advertising. In addition to direct marketing, the network runs ads in tourist magazines showcasing chefs, farmers, and producers and also publishes point-of-interest maps, distributed free of charge to tourist information booths. Money spent by the government to bring farmers and chefs together helps reduce advertising investments and increases the bottom line.

Another interesting idea Roger Clapp spoke with me about is the idea of "foodsheds," a concept similar to watersheds. The idea is to get consumers interested in where their food comes from and to build communities throughout food-producing areas by helping people to understand the interdependence and interconnectedness of farmers, their land, and their communities. Roger went on to explain, "I think that those areas can be defined almost by the individual, but we are looking at regions that might roughly conform to counties, or states, and get people to look into the challenges of feeding themselves and their families from the local community." The idea of "foodsheds" was behind the original development of the Burlington Intervale and the farms on that historic floodplain. Clapp summed up by saying, "We are now working with the farmers there to create a greater understanding within the entire Burlington community, of how their food is grown. Presenting them with the challenge—maybe not of meeting 100 percent of their needs, but a greater percentage of their food needs, from local production, and by paying people a decent wage for producing their food."[52] The more we can help people understand the importance of their food choices and how they impact not only themselves but their children, friends, and neighbors, the more we will change the future of our food supply.

EDUCATIONAL ORGANIZATIONS AND INSTITUTIONS

Chefs Collaborative 2000

Founded in 1993 during an educational program orchestrated by Oldways Preservation and Exchange Trust, Chefs Collaborative 2000 (CC2000) is a growing network of over 1,500 chefs working to promote the chef-farmer connection and educate themselves, their peers, and the general public about sustainability. The Collaborative focuses its attention specifically on encouraging a trend toward regional, seasonal, and whole or minimally processed foods.

Chefs Collaborative 2000 has most recently (October 1998) implemented its "Adopt-a-School" program, which teaches children about sustainability and emphasizes the importance of choosing good, clean, healthy food. Traditions, culture, and history are part of every lesson plan developed by Rick Bayless, Naomi Duguid, Jeffery Alford, K. Dunn Gifford, and Sara Baer-Sinnott. The eight-part program teaches basic food preparations and introduces children between the ages of eight and twelve to foods from around the world, as well as from their region of the country. The classes include geography and history lessons, food tastings, and product identification. Funded in part by American Express, the program is being taught by chefs all over the country. I spoke with Sara Baer-Sinnott, program coordinator, about the students who participated in a class she taught: "They ask such incredible questions and are so perceptive and smart at times that I . . . could tell . . . they really got it. It is a great way to get the students into a culture, having to deal with sounds and language as well as flavors and tastes, like smelling fresh herbs, which some of them have never done."[53] It has been only a year since the first class was taught, but the feedback has been phenomenal—chefs, students, and teachers have much praise for the classes, which really appear to be reaching the children.

Chefs Collaborative 2000 has a number of other programs and benefits for chefs, farmers, producers, and friends. A quarterly newsletter brings information about members, the industry, and food to its active membership. Annual national and numerous regional retreats bring chefs, farmers, and producers together for lively discussions on hard-hitting food-related subjects. Speakers such as Joan Gussow and Alice Waters open discussions on industry challenges, and experts such as Gary Nabhan with Native Seeds/SEARCH, Rebecca Goldberg with the Environmental Defense Fund, and David Wills, consultant to Darden (Red Lobster), bring global issues to the table. The end result for all those involved is a heightened awareness of the industry and an understanding of the questions that need to be asked and how the answers will impact the current and future food supply.

The Culinary Institute of America

Fewer people are cooking for themselves these days. In fact, National Restaurant Association research shows that 50 percent of American food dollars are spent in food service establishments; one out of every four meals is eaten outside the home, and 19 percent of all meals are eaten in cars. Chefs are becoming increasingly responsible for the way America eats. Because the choices we make in our restaurants will ultimately become the choices of our guests who look to us to set an example, it is imperative that we remain well informed about the issues surrounding the sustainability of our food supply. A major challenge for the culinary world is in training the next generation of chefs to understand these complex issues and to help them formulate a plan that will balance peak flavor, financial viability, and the long-term health of the food supply.

I recently sat down with Tim Ryan, executive vice president of the Culinary Institute of America (CIA) in Hyde Park, New York, to discuss the challenges he faces in educating students about sustainability. Men and women arrive at the Culinary Institute with varying levels of previous education and life experience. Some students cross the threshold with a clear vision; others need a great deal of assistance just getting through the basics. Ryan told me, "It is shocking so many students want to be in the culinary arts and do not know if a tomato grows on a tree or is dug up from under the ground like a potato. It is [equally] shocking how many chefs [out there] do not know where things are grown, or have any real recognition that veal comes from a calf. So many people think it comes in cryovac. That is really a shame, and we need to change it." Ryan says that later in the educational program, "We tell our students that sustainable food choices are important for two reasons: one is flavor and the other financial. You get lousy, hothouse-variety produce or things grown in the so-called off-season climates, like Chile, or South America, and they are not as good. [As a consequence] you are serving your customers an inferior product. We are all in the food service industry ... to make good products, and please our guests, but if [your restaurant] is not making a profit, you are not really fulfilling the mission of your restaurant. I ask them, 'Does it make economic sense for you to pay so much more for those lousy, inferior products?' And when you couch it in those terms, you see those lightbulbs of recognition popping on above all the students' heads." On a practical level, as an educational institution, it is difficult for the CIA to operate on a completely seasonal and regional basis, simply because students want to learn to use the largest number of ingredients possible during their time there. If they only have access to root vegetables during a large portion of their classes, they leave with an incomplete education. Ryan stresses the importance of exposing the students to a full range of ingredients, but indicates that the issue of sustainability will continue to be of primary importance to the school.[54]

Eve Felder is associate dean for scheduling and production at the Culinary Institute of America. She is also an alumna. Prior to her return to the CIA as a chef/instructor in 1994, Felder worked as chef at Chez Panisse Café alongside Alice Waters. Shortly after her arrival in Hyde Park, she spent time exploring local farms and visiting with area producers. She told me, "I first became involved with sustainable agriculture and being concerned about food choices [because] I come from a family [living in North Carolina] who were basically hunters and gatherers. ... We always had the best-tasting tomatoes, and the best okra, and my father hunted duck and quail and would catch fish and shrimp. ... I was very used to having very, very fresh ingredients." Because of her upbringing and culinary training, it is natural for her to "forage" for high-quality local ingredients. Her enthusiasm for each and every treasure she brings back from the farms is apparent in her daily classroom lectures. Felder began bringing produce from local farms into her classes to show students the difference between vegetables pulled right out of the ground and the same products after they have traveled 3,000 miles. She told me, "I would go by the farms every morning on my way to work and pick up a flat of Sungold tomatoes, or baby arugula, whatever was fresh, and have my students taste the difference, do comparative tastings. Then I would explain to them that the food has been treated with integrity, that it had been picked, and stored, and gotten to the restaurant in a careful and thoughtful way."[55]

The New England Culinary Institute

Jamie Eisenberg is a chef/instructor at the New England Culinary Institute's Inn at Essex Campus in Essex Junction, Vermont. When I first met Jamie she was the purchasing agent for the inn and was on a panel at the first Vermont Fresh Network conference. As a member of the panel, Jamie was discussing her challenges in purchasing locally for a major culinary school. When I interviewed her, she expressed how important it is to her that the students, as well as the chefs, are instilled with a sense of nature's rhythmic cycle. As purchaser she believed it was her job to make that happen. Jamie first became impassioned about purchasing regionally when Michael Ableman spoke at the New England Culinary Institute (NECI) a few years ago. He so inspired her that she set out immediately to research local purveyors. It was at that time that she discovered Green City Farm at the Intervale in Burlington, Vermont. In collaboration with the farmers, she set to work planning the next year's crops. "I think we had something around $10,000 budgeted for the season. A list of things they could grow, and the things that I wanted, and we negotiated the price based on my bids from distributors, then we upped it by 15 percent. We made a commitment, a signed commitment, to buy vegetables from them, so that they would know that they had a resource for them."

Jamie later discovered that the farm experience was actually the least challenging part of utilizing regional products in a culinary school environment. She had to work to appease the chefs, all of whom wanted specific ingredients, whether or not they were in season, and some of whom were extremely reluctant to implement changes on their menus. Eisenberg was quick to defend the chefs, who already work long hours in a stressful environment. She explained that it was not laziness that prevented them from changing the menus but that the extra time involved in researching local products, understanding the availability, and testing and rewriting menus simply was not there. So it was Jamie who made the extra effort:

> What I have been trying to do [for the chefs] is show them through newsletters, or physically bringing in, what was available from different farms. This way, they do not have to do their own research. And I talk to them a lot. For

example, a chef might come in and ask me for a case of plums, in March, and I'll say, 'Oh I can get them, but I don't want to get them, and this is why. . . .' And I have lots of talks about [how it is not sustainable to get] berries in the winter, and really try to give them alternatives; give them examples about what they can use instead of some of those fresh fruits in the wintertime.

For her part, Jamie does everything she can to get the chefs dried fruits in winter. She has also been known to surprise a chef in the dead of winter with local frozen or dehydrated fruits that she processed in the summer when they were at their peak.

For the students, Eisenberg has integrated sustainability lessons into her classroom lectures. She explains how pesticides that are illegal in the United States find their way across our borders when we insist on buying fresh raspberries in winter from places like South America and Mexico. She began to understand that most students had no idea where their food was coming from, so she started taking them right into the produce cooler, saying, "Let's take a tour around the world now! This came from Guatemala, this came from Nicaragua, this came from Chile, and this came from Mexico!" The students reacted, for the most part, with a mixture of shock and horror. For some hands-on experience, Eisenberg takes students to Green City Farm, where they can get their hands dirty and see the school's produce at various stages of growth. They also have opportunities to meet the farmers and really see what it takes to grow food. Another thing Jamie likes to do is a farmer's luncheon. The menu is preplanned since they know what is in season, but the food is not brought to them; they go to the farm and literally pick their own vegetables fresh from the fields. Then they cook a meal, right there on the farm. The intended lesson serves the dual purpose of enlightening students about seasonality, and demonstrating to the farmers how their produce is being utilized in the kitchen.

Eisenberg also strongly advocates bringing a food preservation curriculum into the school. She believes that preservation is a key component for the sustainability of restaurants in the Northeast and Midwest. Due to Jamie's tireless efforts, some of the chefs at NECI have expressed interest in putting ingredients up for later use, and she is looking to develop an entire class dealing with canning, salting, drying, curing, and sugaring techniques as well as IQF, a process by which foods are individually quick frozen. In the meantime, she has been talking the idea up among the students, explaining that having a pantry, "just like our grandparents used to," is what can carry a kitchen through the winter while helping to keep the purchasing dollars in the community. Her ultimate goal is to see this kind of curriculum happening in all schools—not just culinary schools. She wants to "bring that idea to the kids, and then it will not be such a far-fetched notion to them when they get out there and . . . cook for themselves, or go to the grocery market and buy for themselves. They are not going to see that beautiful, ripe tomato in the middle of January and say, 'That looks great, I am going to buy it,' perhaps instead they will say, 'I want to think about where that came from, and who grew it, and maybe I will buy a bottle of tomato sauce, instead.' One that was produced from tomatoes grown in season."[56] I am sure I am not the only one hoping that Jamie is successful.

Educating the Public

Oldways Preservation and Exchange Trust was founded in 1989 by K. Dunn Gifford, with the goal of "educating opinion leaders about the value of preserving traditional food ways." Gifford and others at Oldways believe that traditional diets with strong cultural ties to specific types of fruits, vegetables, legumes, and grains can be responsible for creating a solid foundation for a sustainable food supply. Greg Dressher, currently director of education at the Culinary Institute of America's Greystone campus, worked with Dunn at Oldways and helped to found CC2000. He explained,

> One piece of [the problem] is how we think about what should be in the center of our plate. The [Western] notion of having ... meat at the center of the plate and grains and vegetables on the side of the plate is not only unhealthy, but it is also unsustainable. And at some point I think we in the United States have to take the part of leading the world in a different direction. If countries like China abandon their traditional approach to a meal, having rice and grains in the center of the plate, and meat as a condiment, we are in for a major agricultural crisis.[57]

It was this kind of thinking that led Gifford to develop a series of "traditional" food pyramids based on the USDA's original conceptualization of the healthy American diet.

With the help of the Harvard School of Public Health, Oldways has produced four traditional diet pyramids, including Mediterranean, Asian, Latin, and vegetarian. An African-American version is currently in progress. Oldways explains, "We have developed a simple graphic to help people visualize the importance of food choices, and suggest ways to make wiser food choices. This graphic ... suggests quite clearly the seamless links among good food, traditions, and health. And it encourages a focus on the pleasures of many ethnic foods ... such as pasta, paella, shish-kebob, prosciutto, sushi, egg rolls, salsas, and tortillas."[58] Oldways' focus on teaching Americans how to eat more healthfully, in conjunction with CC2000's focus on teaching children the same lessons, really begins the process of reeducating the populace about their food choices.

Another great tool for teaching children about their food choices is the school garden, which has been gaining popularity in recent years.

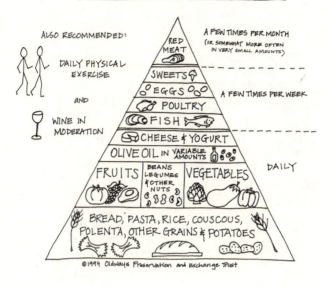

THE TRADITIONAL HEALTHY MEDITERRANEAN DIET PYRAMID

ALSO RECOMMENDED:

DAILY PHYSICAL EXERCISE

AND

WINE IN MODERATION

RED MEAT — A FEW TIMES PER MONTH (OR SOMEWHAT MORE OFTEN IN VERY SMALL AMOUNTS)

SWEETS
EGGS
POULTRY
FISH — A FEW TIMES PER WEEK

CHEESE & YOGURT
OLIVE OIL IN VARIABLE AMOUNTS

FRUITS | BEANS LEGUMES & OTHER NUTS | VEGETABLES — DAILY

BREAD, PASTA, RICE, COUSCOUS, POLENTA, OTHER GRAINS & POTATOES

©1994 Oldways Preservation and Exchange Trust

THE HEALTHY TRADITIONAL LATIN AMERICAN DIET PYRAMID

SIX GLASSES OF WATER A DAY

ALCOHOL IN MODERATION, WITH MEALS

MEAT SWEETS EGGS — OCCASIONALLY or in small quantities

PLANT OILS
FISH MILK PRODUCTS
SHELLFISH POULTRY — DAILY or less

FRUITS | BEANS GRAINS TUBERS NUTS | VEGETABLES — AT EVERY MEAL

DAILY PHYSICAL ACTIVITY

©1996 Oldways Preservation and Exchange Trust

VEGETARIAN DIET PYRAMID

OPTIONAL, OR OCCASIONALLY OR IN SMALL QUANTITIES

OO EGGS & SWEETS

DRINK ENOUGH WATER EVERYDAY FOR GOOD HEALTH

EGG WHITES SOY MILK & DAIRY

NUTS & SEEDS

PLANT OILS

DAILY

BEER, WINE OR OTHER ALCOHOL (OPTIONAL)

WHOLE GRAINS

FRUITS AND VEGETABLES

LEGUMES SOY, BEANS, PEANUTS AND OTHER LEGUMES

AT EVERY MEAL

DAILY PHYSICAL ACTIVITY

©1997 Oldways Preservation and Exchange Trust

THE TRADITIONAL HEALTHY ASIAN DIET PYRAMID

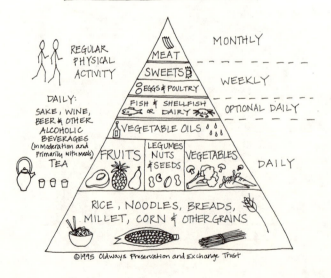

REGULAR PHYSICAL ACTIVITY

MEAT — MONTHLY

SWEETS

EGGS & POULTRY — WEEKLY

FISH & SHELLFISH OR DAIRY — OPTIONAL DAILY

DAILY: SAKE, WINE, BEER & OTHER ALCOHOLIC BEVERAGES (in moderation and primarily with meals)

TEA

VEGETABLE OILS

FRUITS

LEGUMES NUTS & SEEDS

VEGETABLES

DAILY

RICE, NOODLES, BREADS, MILLET, CORN & OTHER GRAINS

©1995 Oldways Preservation and Exchange Trust

Alice Waters is a staunch supporter of the concept. In fact, it was in large part due to her involvement and with the help of the Chez Panisse Foundation that the Edible Schoolyard at Martin Luther King Middle School in Berkeley, California, was created. Students involved in the program grow tomatoes, garlic, kale, lettuce, beans, potatoes, and various other crops on about half an acre of land. They eat some of the vegetables themselves, and some are sold to the public. If Waters has her say, the garden will become the basis for the school lunch program. I had the privilege of spending a morning in the garden and helping the students prepare their lunch for the day, and it was a wonderful experience to see young children arguing over who would get to cook that day. I was moved to see young boys and girls picking out their salad greens and hunting for the perfect cucumber, to watch them turning soil, moving rock, and patching a pond with concrete. I came away with a new regard for the type of motivation growing food and cooking one's own food can provide.

Waters speaks glowingly of how the program can "actually change [the students'] lives when they put their hands in the earth and grow food for themselves and offer it to other people. Gardening connects them to nature, to the earth, to other human beings. It teaches life." The success of the Edible Schoolyard has so influenced the public school system, State Superintendent of Public Instruction Delaine Eastin says, "that within three years, every public school in California that wants a garden will have one." The mission of the Edible Schoolyard is to create and sustain an organic garden and landscape that is wholly integrated into the school's curriculum and lunch program. It involves the student in all aspects of farming the garden—along with preparing, serving, and eating the food—as a means of awakening their senses and encouraging awareness and appreciation of the transformative values of nourishment, community, and stewardship of the land. When I spoke to Alice about how the students perceive the project, she said, "They did a survey this year, where the students had to write down all their favorite classes at the school, including field days and gym— and the Edible Schoolyard project came in third, after gym and field days." By all accounts it seems to be more than fulfilling its mission.

Waters talks about taking the program even further in an effort to truly impact the lives of the students and their families. The Martin Luther King Middle School, under Alice's tutelage, has connected with a local CSA, and has a box of vegetables brought into the classroom every week. She told me that when these boxes are opened in class, it is like show and tell. She explains to the students that the boxes contain what grows at area farms. They get to learn firsthand what the products look and taste like, and some are allowed to take the produce home. Waters said, "Sometimes there was lettuce, and they made a little dressing in the classroom and they ate it. Very, very simply done. But those are the beginnings. Just getting awareness about how important this is. That this is a lot more than just fueling."[59]

The idea that eating is so much more than "fueling" is one that I hope we can get across to Americans around the nation—to the kids in their class-rooms, and especially to chefs in their kitchens, because they are the ones to whom the public looks for guidance.

Stonyfield Farm, best known for its yogurt line, has gone about educating the public in a different way. The company had its beginnings in an organic farming school, founded in the early 1980s by Samuel Kaymen and Gary Hirschberg. In need of funding, the two hatched a plan to make and sell organic yogurt and what began with seven cows has since grown into a $50 million company. In the beginning, Stonyfield's entire product line was organic, but today only 30 percent of it is. When I inquired about the drop in organic production, Mary Jo Viederman, director of "Cow-munications," told me that, due to lack of consumer education, the market demand is just not there yet. She believes, however, that public interest is growing: "Peo-ple are beginning to care about food safety issues, and . . . our goal is to have consumers understand that buying organic is one of the best things that an individual can do because . . . when you buy organic, you are doing some-thing for a sustainable farming system in the future. . . . [One person] can have a great impact."[60] Stonyfield, in spite of an admittedly limited adver-tising budget, has come up with a number of ways to educate the general public. Yogurt lids, normally reserved for company names and product fla-vors, are Stonyfield's billboards for issues like rBGH (recombinant bovine growth hormone) and global warming. The *Moosletter*, or *Moos from the Farm*, talks about life on a small New England dairy farm. Finally, their "Have a Cow" program gives people an opportunity to adopt a New Eng-land dairy cow, and promotes public awareness about the changing face of today's dairy industry.

Mothers & Others for a Livable Planet was initially born out of a National Resources Defense Council (NRDC) project on how pesticides in our food impact the health of our children. Group founders issued a report in 1989 entitled "Intolerable Risk: Pesticides in Our Children's Food," and were later thrust into the spotlight during the public scare over the chemical Alar when cofounder Meryl Streep launched a national public awareness campaign. Wendy Gordon, also cofounder, remembers being swamped with phone calls after the report was made public. Though the NRDC's intent was to enact protective legislation, mothers all over the country were in a panic about their children's safety and simply wanted to know which foods could safely be given to their children—the legal issues were secondary to them. In 1992, after several years of uncertainty about how to best serve the pub-lic, Mothers & Others broke off from the NRDC to carry on a mission to educate the public and shape marketplace activity. Their current member-ship, in excess of 30,000, is kept informed through monthly mailings of the organization's newsletter, the *Green Guide*, which reports the latest environmental and health news along with food safety facts that help readers make informed choices about their purchases. Additionally, Mothers & Others has published several books on the subject of food safety, including *The Way We Grow: Good-Sense Solutions for Protecting our Families from Pesticides in Food* (Berkeley Books, 1993).

Mothers & Others is also currently involved in several projects whose goal it is to mobilize consumers to vote with their food dollars, in support of farmers, for a safe, sustainable, and healthful food supply. The Shoppers' Campaign for Better Food Choices "aims to build demand for a better quality food system that will provide healthy, safe, and affordable food grown in an environmentally responsible manner; to open the marketplace to make it more responsive to consumer concerns and needs; and to create market opportunities for regional, sustainably produced foods."[61] The campaign was launched in 1993, and so far it appears to be working. One does not have to look far from the group's headquarters, in New York City, to see the impact. D'Agostino, a local, sixty-unit, family-owned grocery store chain, was persuaded by Mothers & Others to undertake a new program of "sustainable" foods in all their stores. Guided by Mothers & Others, D'Agostino developed a line of organic, natural, and whole foods they call Earth Goods. The product line was launched in the fall of 1995 along with full-scale promotional materials, including a special circular and a newsletter describing Earth Goods and profiling certain specific products. Another project, CORE Values Northeast, endeavors to, as their slogan indicates, "Bring Good Farming Home" by sponsoring a partnership between Mothers & Others and progressive northeastern apple farmers who are striving to "maintain healthy, ecologically balanced growing environments."[62] As part of CORE Values Northeast, the organization has created an "ecolabel" for fruit that is regionally grown by farmers who utilize biointensive integrated pest management (IPM) agricultural techniques. Finally, in an effort to educate children, Mothers & Others has begun a program they call "Labels for Learning," which promises free passes to zoos or museums in exchange for organic food labels.

Innumerable opportunities exist within our society for education about food and agriculture, but the "student body" involved in one California program might be surprising to some. The Garden Project, based in San Francisco, is the brainchild of sheriff's department employee Catherine Sneed. Her students are prisoners. The project began in 1992 as a way of helping inmates, especially women, acquire skills that would be useful upon their release. Sneed said, "I started the Garden Project as a way to give them work experience, and I found that working in the gardens helped them in a therapeutic way. We began by growing food that we gave to soup kitchens, and we now sell some of the food to restaurants and farmers' markets, but

Share Our Strength (SOS)

Share Our Strength, one of the nation's leading antihunger, antipoverty organizations, began in 1984 when it brought together just a handful of chefs to cook for fund-raising events. Today they mobilize thousands of individuals in the culinary industry to organize events, host dinners, teach cooking and nutrition classes to low-income families, and serve as antihunger advocates. As a result, they have distributed more than $50 million since 1984 to more than 1,000 local, state, national, and international organizations working to prevent the causes and consequences of hunger and poverty. Their philosophy: "It takes more than food to fight hunger. We believe it takes both short- and long-term solutions, such as food assistance, job training, economic development programs and advocacy, to have a lasting impact. We also believe it takes each one of us, sharing our strength, to make a difference." SOS believes that both hunger and poverty are issues that must be addressed. Their research suggests that almost 14 percent of the U.S. population—approximately 36.5 million people—live below the poverty level, and that 40 percent of poor Americans are children.[1] There is more poverty in rural American than in the cities, 40 percent of all homeless people are children, and 7.1 million children live in poor communities—pretty staggering statistics. This problem is vividly brought home when we realize that the U.S. poverty rate is twice the average rate of other industrialized countries.

I have had the good fortune of being involved with a number of SOS events, all of which have been truly enriching experiences for me, my colleagues, and my staff. This year I cotaught an Operation Frontline class with Norman Levitz, who works with me as executive chef of the Putney Inn. These classes form SOS's food and nutrition education program; they use volunteer culinary professionals to teach six-week cooking, nutrition, and food-budgeting classes to low-income individuals across the country. The class we taught was at a church in rural Vermont, where we cooked and shared food with twelve women who were part of the Special Supplemental Nutrition Program for Women, Infants and Children (WIC). Nationwide the program has reached more than 9,000 men, women, and children in eighty communities and twelve states. I have also participated in SOS's Taste the Nation in Burlington, Vermont, which helps to fund the Farm Share program and gives low-income Vermonters the opportunity to participate in a CSA. Taste the Nation, a national program sponsored by SOS, American Express, and Calphalon, is the nation's largest culinary benefit to fight hunger. Cities across the country hold events at which 100 percent of ticket sales goes to antihunger and antipoverty efforts. By its tenth anniversary in 1997, Taste the Nation events had raised more than $27.8 million to fight hunger.

1 Share Our Strength, "Hunger and Poverty in the United States, U.S. Census Bureau 1997," Share Our Strength, 12 December 1999, http://www.strength.org/hunger/poverty.html.

we continue to give food to senior centers, homeless kids, and soup kitchens."[63] The project's main emphasis is on employment training. Upon their release, ex-inmates may be employed by the project and are paid between $7.15 and $8.00 per hour, which helps them smooth the transition into their "new" lives. Since its inception, the Garden Project has employed approximately 1,200 people, and the jail program has utilized more than 10,000 prisoners, all of whom were successfully reacquainted with the origins of their food supplies, and many of whom discovered a connection that helped change their lives forever.

LAND TRUSTS: KEEPING FARMERS FARMING

Keeping farmers farming is certainly part of a sustainable agriculture system. The Vermont Land Trust has been functioning as a means of preserving farmland in the form of conservation easements since 1977. It began in an effort to save some land in the Woodstock, Vermont, area, but quickly grew to become a statewide organization with an agenda that blends the conservation of land with the viability and sustainability of local agriculture. Darby Bradley, president of the Vermont Land Trust, and I spoke about the Land Trust, which has had a tremendous impact on state agricultural land. He explained that beginning around 1987, "[The Land Trust] really began to have a significant number of farm projects, and the farm community was quite suspicious of the idea of selling development rights. There was concern that the Land Trust would tell them how to run their farms, or that the banks would not make a loan, or they could not sell their property with a conservation easement on it. In the last six or seven years, we have gotten over those hurdles, and we are now conserving about thirty farms a year."[64] Thus far, approximately 220 farms covering nearly 50,000 acres of land have been protected. That accounts for about 12 percent of all Vermont farms. I asked Darby what exactly a conservation easement was, who paid for it, and how the money was eventually utilized. He told me that the money for the trust comes from three sources: the Housing Conservation Trust, federal and state grants, and private charitable organizations and trusts. Funds received by the farmers are most often used to pay off long-term debt, to purchase or upgrade equipment, or to actually buy the land. The primary goal is to keep the land from being developed. It must, by right of the easement, always be agricultural land or open space. The Land Trust's secondary goal is to keep the farms economically viable. It was with that in mind that they sponsored a "Taste Vermont" event in 1997, which paired Vermont chefs with farmers from the conserved lands in an effort to encourage personal relationships as well as promote regional farming.

David Batcheldor's family has been farming in Stratham, New Hampshire, for over 100 years. In 1991 he sold the development rights of his farm to the state and town in an effort to add to the farm infrastructure (machinery, equipment, and farm buildings) and make his organic farm, at that time the largest certified organic farm in the state, more viable. He explained that no one, not even the town, has the right to develop the land, and that by law it can only be utilized for agricultural purposes. Purchase of the easement created an influx of cash flow to the farm, which apart from the change in potential development rights continues to operate just as it always has. In essence, barring the passage of legislation, the easement will remain in effect forever. I asked David how he felt about "selling" the farm that has been in his family for generations. He put it this way: "We were a little concerned when we were first thinking about it, wondering if it would be possible, in thirty or fifty years, to [maintain a] farm here in Stratham, because of the population growth. Since that time it has become more and more clear that it is very important that this stay a farm, whether I have anything to do with it or not. I just think it is critically important that this land continue to be farmed in some manner."[65] I have known David and his family for over twenty-five years and visited the farm numerous times. During my last visit I walked the fields, helped harvest some produce, stocked the farm store, and shared food in their kitchen with the family. My hope is that programs like the Land Trust will help family farms remain part of our future and keep them from becoming part of our past.

IS SUSTAINABLE SUSTAINABLE?

> Without question, the most important and most often-debated principle is carrying capacity. What is the rate at which and manner in which the world can sustain the human population that exists and is growing? We do not know the answer to that question yet. In all ecosystems, the availability of food and nutrients becomes the ultimate arbiter of population size. But we are humans, not fish in a pond.
>
> —Paul Hawken, *The Ecology of Commerce*

In his book *State of the World 1998*, Lester Brown stated, "Feeding 80 million more people each year means expanding the grain harvest by 26 million tons, or 71,000 tons a day. Not only is the world now adding 80 million people annually, but it is projected to add nearly this number for the next few decades, reaching 9.4 billion in 2050."[66] Irena Chalmers, an outspoken advocate for sustainable food systems, reinforced Brown's assessment: "In the next two to four generations, world agriculture will be called on to grow as much food as has been produced in the entire 12,000-year history of agriculture."[67] Most agricultural economists agree that by the year 2050, farmers will have to produce nearly three times as much food and fiber as they do today, and many have their doubts about sustainable agriculture's ability to meet those needs. Their uncertainty stems in part from the notion that sustainable agriculture generally requires more land to produce the same amount of product as conventional agriculture (although John Jeavons argues against this point). What we need is more product from less land. In *Farming to Sustain the Environment,* Denis and Alex Avery write, "The earth will not be providing any more land over the next 50 years, and as many sustainable farming systems stand today, they are 50 percent as productive as current conventional practices. Even under the best of circumstances we will need more land to produce as much food as we will need in 50 years—some estimate that we will need as much as six times the current amount in production."[68] Without changes in agricultural practice, we will surely destroy the land and water, and possibly even plant and animal biodiversity; however, if we are successful in making the change from conventional to sustainable agriculture, we risk the greatest famine the world has ever seen.

It is very difficult for Americans, and indeed the citizens of most Western nations, to comprehend the enormity of famine, since on the whole we are very well fed. We are also extremely unwilling to acknowledge the impact of our own selfishness in requiring such diversity in our food supply. How will population growth affect us here in America? It seems clear that emerging nations that continue to experience wild increases in population will have continued food deficits. Countries such as the United States, with its annual food surpluses, will be cast in the role of provider. As our population increases and requires more of our own production, we will be forced to make difficult decisions that will call into question our sense of morality and justice. In *Losing Ground,* Erik Eckholm puts it this way:

> Given the past economic record and foreseeable economic future of many of the poorest countries, a good share of the potential food gap will probably be left unfilled by the commercial market. If so, scarcity will manifest itself in a rising incidence of malnutrition and premature death, the common assumption of steady historical progress toward a better life for all, shattered. Under these circumstances the more affluent countries, particularly those with a food surplus, will face choices and responsibilities so politically sensitive that they may not be able to deal with them rationally. What portion of the exportable food should be reserved for charity, what portion for cash customers? Should domestic consumers alter their diets to make food available for the impoverished abroad? Are food gifts to needy countries moral or even responsible if they encourage greater tragedy in the future?[69]

Is it our responsibility to take care of the rest of the world's people? Some wonder how we can even consider the plight of those in other countries when we are not doing an adequate job of caring for our own citizens. Senator Patrick Leahy recently said to me, "Look at the United States. It is the only major power on earth [with] the ability to raise all the food we would need for a quarter of a billion people and have enough food left over to export enough for tens of millions more. And yet we have a lot of people, especially children, who go hungry, children who often go to school inadequately fed. We have pregnant women not getting proper nutrition, and neither are their children afterward." Second Harvest, one of the nation's largest food assistance organizations, stated in a recent study that it provided food for over 21 million people in 1997.[70] Included in this startling statistic are 8 million children, and more than 3.5 million elderly.[71]

If those figures are correct, distribution is very clearly an issue. The World Health Organization estimates that one-third of the world's population suffer from some form of malnutrition that in turn leads to high infant and child mortality, impaired physical and mental development, and impaired immune systems. "Production is not the problem," said Guido de Marco, president of the UN General Assembly, in 1990. "Distribution, transportation, storage, and guaranteed access on a regular basis to suitable food—these are the principal obstacles which need to be overcome. These are the most urgent problems and should be treated as such."[72] If he is right, what are we to do about these problems?

As we end this section and move on to the next chapter, we continue to be faced with complex issues of great magnitude. How can we produce a safe food supply, assure adequate and safe water sources, feed an ever-burgeoning population, and do it all sustainably? Chapter 3 looks at how agribusiness's actions on these issues have shaped public policy and ultimately our food supplies. Legislation, subsidies, international policies—all often dictated by agribusiness—have become part of the web that supports, distributes, and supplies America's food.

▪ 3 ▪

Agribusiness:
Controlling Our Food

There seem to be but three ways for a nation to acquire wealth.
The first is by war, as the Romans did, in plundering their con-
quered neighbors. This is robbery. The second by commerce, which
is generally cheating. The third by agriculture, the only honest
way, wherein man receives a real increase of the seed thrown into
the ground, in a kind of continual miracle, wrought by the hand
of God in his favor, as a reward for his innocent life and his vir-
tuous industry.

—Benjamin Franklin, *Positions to be Examined*
Concerning National Wealth

In spite of the fact that farms have been reincarnated under the corporate
umbrella, most Americans would probably admit that when they hear the
word *farm*, they still think of small fields filled with wheat and corn, a few
cows, some pigs and chickens, and a big red barn. In reality, that is about as
far from the truth as it could be. Today's industrialized farms inhabit a flattened landscape, dotted not with trees, farmhouses, animals, and farmers
whose backs are bent in service to the earth but with huge motorized vehicles that plant, water, and harvest virtually unaided by human hands.
Animals do not lazily graze these fields; they stand shoulder to shoulder,
practically immobilized, fed by computer, eating and sleeping away the
hours until their inevitable slaughter. The land becomes poisoned; the animals are robbed of a stress-free life and of their dignity.

It is difficult to imagine a human connection to these vast, yield-driven
farms that leave our food strikingly soulless. Where do we chefs and consumers fit into these new and anonymous systems? How do they affect our
food memories and the way we provide sustenance for our friends and families? The size and scope of corporate farming automatically removes the
element of nurture that exists in traditional, small-scale farming—both for
the products and for the consumer. What remains is the business of farming—or agribusiness.

AGRIBUSINESS—AN UNINVITED DINNER GUEST?

Agribusiness has everything to do with the food we eat every day. In fact, on the day you read this book, chances are good that you will eat food produced by an agribusiness company. In the late 1990s, the majority of America's food supply is being produced by only four companies that "control 87% of all beef packing, 74% of wet corn milling, 73% of sheep slaughter, 60% of pork packing, and 55% of broiler [chicken] production. [Additionally,] five companies control 76% of soybean crushing, 62% of flour milling, and 57% of dry corn milling."[1] ConAgra, Cargill, and Archer Daniels Midland (also known as ADM, "Supermarket to the World") are food producers whose corporate names and logos are strongly tethered in the consciousness of the American public to our food supply. However, a company like Philip Morris, to whom 10 cents on every dollar spent in a supermarket goes, is more likely to conjure images of the Marlboro Man than its food holdings, which include Miller Brewing, 7-Up, and General Foods (Post Cereals, Maxwell House Coffee, Sanka, Jell-O, Kool-Aid, Oscar Mayer, and Log Cabin Syrup), to name a few. In addition, a small minority of "brand name" companies control huge majorities of their specific markets. For instance, Campbell's alone is responsible for 76.5 percent of the canned soup market, Coca-Cola and Pepsi together control over 74 percent of the soft drink market, and 84 percent of cereal is produced by Kellogg's, General Mills, Post, and Quaker Oats.

These companies have tremendous buying power, simply because of their size. They can purchase millions of dollars of product at the lowest possible prices, and are in a position to set industry standards. Chemicals, pesticides, herbicides, irradiation, bioengineering, food additives, production techniques, and land usage are directly related to the types and quantities of items these companies decide to buy and sell. Their participation in purchasing or utilizing these products and technologies paves the way for the future of food production and influences consumer choices as well as the ultimate health and safety of our food supply. The top ten "farms" in the United States today are international agribusiness corporations with names like Tyson Foods, ConAgra, Gold Kist, Continental Grain, Perdue Farms, Pilgrim's Pride, and Cargill—each with annual farm product sales ranging from $310 million to $1.7 billion. This concentration of power is largely related to the 4,100 food industry mergers and leveraged buyouts that occurred in the United States between 1982 and 1990. The shifting distribution of power continues as we head into the twenty-first century.

Hybrid seed companies have occupied their place at the beginning of the American food chain since the early 1900s. Until recently, a large number of companies shared control over a wide variety of crop seeds. Today, however, ownership of seeds, or germplast, has been heavily consolidated.

Throughout the last decade of the twentieth century, chemical companies and industrial giants such as Monsanto, DuPont, and Cargill have virtually become household names. In fact, "more than 48 percent of all nonhybrid seeds are available from only one source. A single corporation controls 66 percent of the world's banana germplasm, two companies control 43 percent of all barley, four companies control 79 percent of all bean varieties and six companies control 66 percent of all lettuce types."[2]

All this consolidation of power in the agricultural sector has had major impacts on the industry. Not only are these new megacorporations able to heavily influence market demand, but they can also grow food plants and animals at much lower prices than a small farmer might be able to do. As a consequence, the market may be forced to continue consolidation as well as increase chemical usage (to increase production) just to keep up financially. Between 1979 and 1997, the producer share of retail beef revenues dropped from 64 to 49 percent and the farm price for slaughter steer fell 50 percent. It is much less expensive to produce our food today than at any other time in history.

One would think that consumer costs would adequately reflect the reduction in production costs; but in fact, actual public purchasing prices dropped by only 15 percent during that same timeframe.[3] This restructuring of agriculture has contributed to decreasing the farmers' share of consumers' food dollars from 41 percent in 1910 to 9 percent in 1990.[4] As a result, from 1935 to 1989, the number of small farms in the United States declined from 6.8 million to under 2.1 million; during the same period, the U.S. population roughly doubled. Along with the failure of so many small farms, the United States has seen a dramatic loss in the number of small businesses, including local suppliers, implement dealers, and others that once supported the small farmer. Entire rural communities have disappeared. Meanwhile, the agribusiness corporations have grown and consolidated their power.

With sales topping $110 billion, the fast food business makes up a significant segment of the food industry. Its anonymous drive-up windows stand like sentinels at highway's edge, quietly, ubiquitously linking every city and town in America. Some of the country's most prime real estate is given over to these brightly painted cinderblock totems of dispatch and parsimoniousness. Golden arches promise hamburgers, fries, and Cokes for all; gigantic red and white buckets are tirelessly and obligingly filled with crispy, greasy fried chicken; and taco after taco streams into the street like music from a wild and chaotic carillon. Have you ever considered for one moment what kind of stress that segment of the industry alone puts on the resources of our planet? McDonald's and Burger King together purchase approximately 2 million pounds of hamburger a day. Two million pounds! To grow one pound of beef requires seven pounds of feed, which in turn requires 175 gallons of water. These two burger chains, just for their hamburgers, use 350 million gallons of water and 14 million pounds of feed—a day. That is only a fraction of what goes on behind the scenes in fast food, and it is only one day of one month of one year.

Fast food, like agribusiness, strives for uniformity. The goal, for both industries, is consistency of products. They work together to create "flawless" fruits and vegetables—even animals—weeding out the less desirable strains. They demand specific sizes, weights, levels of ripeness, and extended shelf lives of various products, which forces farmers to grow only specific varieties, and ultimately narrows the gene pool. In fact, Greenpeace reports that by the year 2000 we will have lost 95 percent of the genetic diversity that was used in agriculture at the beginning of the century. The fast food industry, because of its need for inexpensive beef, also has a devastating effect on tropical rain forests around the world, which are frequently burned and cleared for use as pastureland. The effects of this unprecedented loss of rain forests may be felt for decades, if not centuries, to come.

Some of the forces behind the fast food industry, including agribusiness, are at the forefront of technological development. Agribusiness funds 20 percent of the nation's agricultural research, and its leaders believe they have found the answer to their need for consistent products in biotechnology, a growing industry that espouses its ability to provide uniform, cost-effective, high-quality product in an efficient and highly profitable manner.

BIOENGINEERING, BIOTECHNOLOGY, AND OUR FOOD

From an agricultural standpoint, contemporary biotechnology and bioengineering have been focused on producing "better" fruits and vegetables. Agribusiness generally defines "better" as a plant that requires less water to grow, whose fruit is easily harvested, and which is adaptable to climate as well as excessive chemical application. Forerunners in these types of technological developments include Monsanto and DuPont (along with their subsidiaries). Both companies got their start in the industry as producers of chemicals or chemically treated products, such as carpets designed to resist stains, and both have undergone a conscious and calculating mutation, redirecting their corporate focus to become "life science" specialty companies. Their aim is to produce a "better" food supply through technological advancement, and they are spending billions of dollars, according to the *Wall Street Journal*, in a race to "rewire the nation's crop of corn, soybeans, and other mainstays for use in everything from new types of food to pharmaceuticals and plastic."[5]

In May 1998 the *Washington Post* reported, "Through direct investment and alliances, the two rivals [Monsanto and DuPont] are getting control of seed producers for most major U.S. crops. Combined, they would control roughly half of the U.S. seed market for soybeans and even more of the seed market for corn, the nation's two largest crops. Monsanto alone stands to control a staggering 80 percent of the U.S. cottonseed market, if pending transactions win regulatory approval." Eventually these two mega-giants hope to take orders for new types of crops and "pop" them out of a test tube for companies like ConAgra, Nestle, or Philip Morris.[6] The "science" that allows for these innovations is a new one, the basis of which is DNA. The technology, which is still on the uphill side of the learning curve, allows genes from one plant or animal to be isolated and introduced to that of another. Although the technology is still in its infancy, multinational companies around the globe are engaging in unprecedented mergers and acquisitions, staking entire fortunes and futures on it. Board members at these corporations believe that genetic engineering will be as important to the economy and the buying public as oil has been. Whoever controls germplasm, theoretically, controls the future of our food supply. Chemical and pharmaceutical companies, the first in line to conquer the technology, have already asserted themselves at patent offices around the world. They are positioning their strategic interests in order to make decisions about what we eat—from how it is grown to its nutritional value. In the creation of a new food supply, agribusiness will also be in a position to make choices that will have a dramatic impact on the environment.

LOSS OF BIODIVERSITY

Since the beginnings of agriculture, farmers have worked to preserve food plants and animals that best suit their needs and those of the consumer. At first, this simply happened through natural selection—weaker plants and animals just did not survive. Later, farmers learned to save seeds from their hardiest, most productive plants for the following season. As time passed, agronomists came to understand the nuances of animal cross-breeding and plant hybridization. These "new" plants would have the desired traits from both parents, but they would not carry the traits of both into the next generation—they would reproduce but would not replicate exactly what had come before.

By making conscious choices to breed out certain plants or animals, we have placed severe limitations on the biodiversity of plant and animal life on Earth. Until now, the process has been long and slow. The shift from hybridization to bioengineering is actually an enormous leap. While hybridization utilizes traditional plant-breeding techniques, bioengineering literally takes genes, in the form of DNA, from plants or animals and implants them into a completely different plant or animal, creating not a new variety of the same life form but an entirely different entity. The implementation of this type of advanced science by large corporations could increase the loss of biodiversity dramatically, as plant varieties and species are tossed aside by those who have the ultimate control over the seeds and animals farmers are buying. These industrial behemoths spend a considerable

amount of time and money researching which plants will be the most lucrative in the marketplace, and then working to "invent" the ones that suit
their needs. In *Rain Forest in Your Kitchen* Martin Teitel reports that

> [agribusiness] policies are leading to a dramatic narrowing of the genetic
> diversity of the planet and creating an environmental and economic crisis
> of unparalleled dimensions. By relying on only a handful of "market-effi-
> cient" varieties and breeds for production and marketing around the world,
> these corporations have greatly diminished the number of plant and animal
> species under domestication. The human population of the planet ... may
> well find their food choices dangerously narrowed to meet the requirements
> of mass production, economies of scale, marketing prerequisites, and mar-
> ket performance. If biological diversity continues to diminish at the present
> rate, future generations may find themselves without sufficient genetic
> reserves to sustain their own survival.[7]

To complicate matters, this technology and research is expensive. No one
wants to invest hundreds of millions of dollars developing products that
can be created and sold by anyone. The government, in an effort to promote and encourage this groundbreaking scientific study, has recently
stepped in to allow the patenting of new bioengineered life forms.

Allowing corporations to patent food plants creates all sorts of complications in the food supply—not the least of which is that farmers are not
being permitted to save seeds from one harvest to the next, as they have
traditionally done. When they purchase seeds from companies such as Monsanto, they are required to sign contracts declaring that they have no
intention of saving seeds for their own use or for sale to other farmers. In
addition, when a farmer buys bioengineered seeds, he is now being required
to pay to the creator of the germplast a "technology tax" over and above
the cost of the seed for every acre planted. With this arrangement, agribusiness has everything to gain, while the small farmer may be forced into
financial ruin. Many farmers see seed saving as their God-given right and
have quietly challenged Monsanto and other companies like it by flouting
their contracts and saving seeds anyway. In response, Monsanto had over
400 lawsuits pending by the end of 1998 and is actively pursuing remediation from farmers who they believe have "stolen" their seeds. In a press
release that same year, the company reported the findings of one of the first
court cases that it settled:

> Following a recent seed piracy investigation, David Chaney of Reed, Ky.,
> admitted to illegally saving and replanting Roundup Ready® soybeans.
> Chaney also acknowledged that in return for other goods, he illegally traded
> the pirated seed with neighbors and an area seed cleaner for the purpose of
> replanting. All of those involved were implicated when Monsanto made the

discovery. Chaney's settlement agreement terms include a $35,000 royalty payment as well as full documentation confirming the disposal of his unlawful soybean crop.[8]

While I sympathize with Monsanto's right to reap financial return for its products and technologies, I also firmly believe that if companies are permitted to own our food supplies, their power becomes tantamount to the power to control our future.

ROUNDUP READY, THE TERMINATOR GENE, AND rBGH

Corn, wheat, potatoes, tomatoes, and soybeans are but a few of the plants that now have bioengineered cousins. Monsanto, in particular, has been a leader in developing these plants, and one of its most recent publicly debated inventions is a line of Roundup Ready plants, whose germplasm has been modified to withstand the application of significant quantities of Monsanto's Roundup chemical herbicide. This means that Roundup can be sprayed, often numerous times throughout the growing season, without fear of destroying the crop. As with most other controversial scientific developments, there are both positive and negative implications to this technology. On the positive side, Roundup Ready crops enable farmers to practice "no-till" agriculture, which means the soil is left intact and soil erosion is kept to a minimum. This is one of the ways in which Monsanto portrays itself to be striving for sustainability. Conversely, the introduction, and possibly heavy application, of an herbicide like Roundup practically guarantees the destruction of every living plant within the spray zone. Additional problems may arise through loss of habitat for beneficial insects as well as overspraying into fields whose plants are not modified with the Roundup Ready gene. If the herbicide is overused and enters the groundwater system in high concentrations, it may end up in our drinking water and could have a negative impact on the small birds and wildlife in the affected area. At present, Monsanto denies its product's potential for severe harm and strongly claims to be working toward a sustainable tomorrow by protecting the future of the planet and protecting our food supplies.

Altruism is not necessarily the only goal behind Monsanto's business plan. Roundup, currently one of America's most popular herbicides, will be out of patent protection in the next decade. One way of protecting company profits is to produce another product that is dependent upon the first. By locking farmers into using their newly developed seed, Monsanto also requires them to continue to buy the herbicide. If Monsanto can convince farmers its products yield a better, more profitable harvest, they will gladly patronize the corporate giant.

In the 1990s Monsanto spent hundreds of millions of dollars acquiring seed companies, "inventing" plants, marketing its new creations to the agribusiness community, and working to protect its investments. Roundup Ready seeds are one way of doing that.

The "Terminator" gene, originally named the Technology Protection System, is another way. Developed in a collaborative effort between the USDA and Delta Pine & Land, the Terminator gene's primary purpose is to render a plant sterile. In essence, once implanted, the gene stops the plant from producing viable seed. In all, we provided $190,000 tax dollars (combined with Delta Pine's $530,000) to fund this revolutionary research. The government's rationale behind the investment was that companies would cease to invest in future technologies if they could not be assured of a return on investment. USDA spokesman Willard Phelps stated, "The new technology will be so widely developed that future farmers will be able to purchase only sterile seed."[9] Henceforth the seed-saving issue, lawsuits and all, ceases to exist. Joan Gussow described her feelings about the Terminator gene to me: "In my view, that is a fist in the eye of God, and I am not even religious. It is so outrageous, the idea that you are going to build sterility into seeds, which are the basis of all life, it is so frightening. We are talking about building sterility into the second generation, so that the seeds will not germinate at all. That is really a sort of wildly dangerous thing to do."[10] Delta Pine calls this technology "Control of Plant Gene Expression," and under U.S. law, since DP&L worked with the USDA to develop the technology, the company has the option to negotiate an exclusive license. Martha L. Crouch, associate professor of biology at Indiana University, explains how the effect of the Terminator gene differs from hybridization:

> When the Terminator gene is part of a plant whose seed is saved and which is replanted, the second generation is killed. With hybridization, the second generation is variable, but alive. Any genes present in the hybrid will be present in the second generation, although in unpredictable combinations. Therefore a plant breeder who wanted to use the genetic material from the hybrid in his or her own breeding program could retrieve it from these plants. With Terminator, the special genes, such as those for herbicide tolerance, would not be easily available for use by competitors.[11]

Profit is clearly the motive. The December 1998 edition of *Acres U.S.A.— A Voice for Eco-Agriculture* reported that even government employees stand to gain from money earned by this technological advance. "When the Terminator seed hits the market, in an estimated five years, the USDA is expected to earn royalties of 5 percent of net sales and the USDA scientists will personally receive a portion of the profit."[12] I believe that this government funding, as well as profiting by our nation's civil servants, sets a dangerous precedent for the future. With that kind of involvement one is hard pressed to find a disinterested party whose focus is on the best interests of farmers and the consuming public. The USDA intends to make

Terminator technology available to all seed companies, who anxiously await its arrival, as early as 2004. The idea of a Terminator gene may be hard to grasp, but the insinuation that we will soon invent a nonreproducible food is virtually incomprehensible. The system would make seed saving and sharing impossible, and give control of our food supplies to a handful of megacorporations who could then charge what they want for the seeds and virtually add or subtract foodstuffs from the marketplace as profits dictate. This would be a truly frightening new world order. At the close of 1999, due in part to negative publicity and consumer outcry, Monsanto stated that it will not utilize Terminator technology in the foreseeable future. The government has not yet made a similar commitment.

In the course of my research, I thought it important, as well as fair, to speak with someone at Monsanto about the company's views on the products it creates. I interviewed Phil Angel, former vice president of marketing, in an effort to shed some light on the company's perspective. He told me,

> There are four inexorable, unavoidable forces shaping our future. One is population. We are 5.8 billion now.... In 2025 we will have somewhere around 8-plus billion people in the world. That is there. That is unavoidable. Two: the amount of arable land available to produce food in this world is about where it is, and it is not going to get appreciably larger—6 million square miles. You can expand it, but you expand it at a price, and the price is wetlands, rain forests, and the like. So you are looking at a resource base that produces what we eat today, that is not going to significantly change in its size.... In the future we are asking that resource base to essentially provide somewhere between 60 to 100 percent more than what it produces today to meet food demands in the future. Three: the world's population is getting wealthier. We are developing wealth at the rate of 3.4 percent a year, which means you are going to double the world's economy in twenty years. Growing middle classes in the developing world make bigger demands on a whole range of resources, including food. That is the problem with population growth, that 90 percent of the growth in the next twenty-five years is going to take place in the developing world, where they are increasingly unwilling to accept a living standard that is where it is now. They want a piece of the economic pie, and they are going to get it, one way or the other.

Angel went on to explain that Monsanto is striving toward creating products that encourage sustainable agriculture:

> I am fascinated by the idea of sustainable agriculture, because I believe that the issue of how you provide food to an increased population from a limited resource base is the classic definition of the challenge that was presented in the Bruntland Report in 1987.[13] [We] have a fixed resource that [we] know [we] cannot destroy, because it essentially provides life, it is what we

use to grow our food . . . but [we] are going to have to find a way of making that land more productive, and there are going to be a lot of ways to do that. Protecting the land by finding ways to cut back erosion [using] no-till farming is clearly the most significant change that . . . can be made in the way we grow our food. Biotechnology in agriculture is another . . . important tool that is increasingly being recognized by aid bodies and the World Bank, as a way of increasing the yield of food supplies. At the same time, bioengineered products potentially reduce the amount of outside inputs like fertilizer, pesticides, and herbicides [and can help increase yields]. . . . The expectation is that in the future [we] are going to be able to create crops that will give you even more yield per unit of land—all for less cost and less environmental harm. That is where we see ourselves in this whole equation, not the answer, but an important part of where the world is going to have to go to meet [future] demands.[14]

Senator Patrick Leahy has been involved with environmental and sustainability issues for many years, and while he is concerned about the power these companies have he also believes that development of bioengineered plants will continue, and the industry will expand. He told me that he "can see bioengineering that will allow some plants to be grown in poorer countries that will not require pesticides, and will not require a lot of chemicals for fertilizers." He continued, "I do see a future for some bioengineered strains that are resistant to indigenous pests or weather conditions,"[15] but cautioned that these types of products are generally prohibitive in cost developing nations anyway. I believe that it is possible for bioengineered plants to play a role in eradicating world hunger and that they may help to reverse the overwhelming loss of the world's topsoil, but like Senator Leahy, I believe that bioengineering requires a great deal of scrutiny and debate. We must find a way to harness the power of these emerging technologies. As consumers, we need to ensure that there are strict regulations and testing. The entire industry should be open to intense public dialogue, debate, and oversight.

Bioengineering has both opponents and proponents, the sizes of the various groups often dependent upon the group demographic. The dairy industry's battle over rBGH, a bioengineered milk production enhancer, is a good example of how the introduction of these products provokes unexpected marketplace reaction. Cattle normally produce hormones that naturally regulate growth and milk production. Monsanto recently discovered a way to bioengineer a replica hormone, known as rBGH (recombinant bovine growth hormone), that can be administered to cows by dairy farmers to enhance, or increase, milk output. It appeared on the market amid controversy in 1994 and is currently being utilized by numerous commercial U.S. dairy operations. rBGH does indeed enhance milk production; however, many believe that in the United States we have no need to increase our milk supply and therefore should not be utilizing the growth hormone at all. Michael Greger, author of "Bovine Growth Hormone: What You May Not Know," believes that "the most persistent economic problem faced by the dairy industry today is overproduction." He says that "for the last ten years [prior to 1998] the dairy surplus has been over a quarter million tons of milk each year and the government uses a billion of our dollars every year to sop it up. According to the US Federal Office of Management and Budget the projected increase in milk production caused by rBGH introduction will cost American taxpayers an additional $116 million of further price supports in 1995 alone."[16] Some experts believe that inevitably this program will cost U.S. taxpayers billions of dollars for federal purchases of the increased surplus of unsold dairy products.

Biotechnology: Chefs Weigh In

I interviewed chef Rick Bayless, and we discussed both sides of biotechnology at length. He said:

> You know, there are so many good arguments on both sides. I wish that there was a program for wise people in our country, that somehow the great wise ones rose to the top, not the Supreme Court, but people that really knew, that were really sages. We could just turn it over to them, and say, "You guide us, day by day." And all the scientists would bring all their information to them, and say, "This is what we are thinking about doing, is it the right thing?" And they would say, in their great wisdom, whether it was or not.

Rick was so interested in understanding bioengineering that when Monsanto invited him along with several other chefs to fly to Saint Louis for a day tour of the plant and to talk about Monsanto's Flavr Savr tomato, he quickly accepted the invitation. He told me,

> Man, their arguments were so strong! They said, "Cotton is the crop that uses more pesticides and herbicides than any other crop known to man." And there I was sitting in my all-cotton clothes, and they continued, "We want you to feel good about wearing cotton clothes, and we can genetically engineer this cotton to repel all these pests, so that it can go without any pesticides." Wow. That sounds pretty great, you know? Personally, I think the power of bioengineering is so immense, it is not generational, bioengineering opens up the entire world to us, and you can cross one thing with something that it would never cross with any other time. So I think the power of it is so enormous, that in a sense it is the same thing as unleashing nuclear energy. It can be used for some really good things, but it can also be used for some incredibly destructive things. From my standpoint as a chef, I do not want bioengineered food in my restaurant because I am not convinced that the people that are holding that technology have the good of mankind at heart. I have not been convinced of that yet, and because it is such a powerful tool, I just simply want to hold off from it. I do not want to start putting money in their pockets and say, "Wow, you guys are doing a great job."

Biotechnology: Chefs Weigh In (continued)

One of the most interesting parts of Rick's story was that even though he was at the Monsanto plant to evaluate the Flavr Savr tomato, he never got to try it because it had not yet been approved.

> We could not taste it, which was just hysterical, because here we were looking at these things in these boxes, in Plexiglas boxes as though they were specimens from another planet.... We had one of the scientists there who was a pretty level-headed guy, he certainly was not defensive at all.... They basically said, "We just wanted you guys to hear our side of it, so that when you make up your own minds, you will have some good information to work with."[1]

Other chefs, like Michael Romano (Union Square Café), also have a lot to contribute to the debate. Romano told me,

> I think most chefs end up commenting on bioengineering from an emotional level, which is great, we should be passionate people, but we are not scientists. I am not a genetic scientist and so I can only comment about things that I can intuit based on what I have read. To me, in terms of genetic engineering, the jury is still out, but I will say that I am inclined to think that it can be used for good and I am willing to give it the benefit of the doubt. I am not in favor at all, and I do not care what the science says, of interspecies gene-splicing. I do not think God intended flounder genes to be put into tomatoes, that is not part of His plan, nor is it part of mine. However, I do recall from my fifth-grade science class, Gregor Mendel, who messed around with peas and was the father of genetics. Given that we have been doing this for centuries and centuries, if there are new tools that can allow us to do it more efficiently, then I say, "Great," we should be open-minded. If it can be used for the good of mankind, to teach us how to grow crops in the desert, or to reduce our dependence on pesticides by coming up with plant varieties that are naturally pest-resistant, great, what could be wrong with that?[2]

When I spoke to Jesse Cool (Flea Street Café), she told me that she felt that the issue of bioengineering was confusing, but clearly not all wrong.

> I have a problem with the people who say, "It's all bad," I have a lot of mixed feelings. I feel that since Mendel spliced those first peas, we have been playing with genetics, and to the betterment of the world! Clearly part of the reason we have the wonderful food in this country we do, is because people were playing with the food. But, given that we are running out of water—we are running out of everything, really—then how do we balance the fact that genetically engineered food can be made to need less water? Somebody is going to [create] this great strain of corn ... with less water, and no chemicals, and if that is bioengineering, then what? We have a lot to learn, and technology may help us get where we want to go, I think.[3]

1. Rick Bayless, interview by author, Chicago, Ill., 9 July 1998.
2. Michael Romano, interview by author, New York, 6 August 1998.
3. Jesse Cool, interview by author, Menlo Park, Calif., 29 July 1998.

On the other hand, a product such as rBGH could be a blessing for people of emerging nations. In countries where animals and people are both hungry, where land and fodder is scarce, where children die of starvation daily, these innovations may very well be useful. However, there are many challenges to that philosophy. Can emerging countries afford the technology? What will the costs be of training the farmers to understand it? And perhaps the most serious question, would it be easier to ship our surplus milk to countries that needed it? Would this be a reasonable answer to our oversupply of milk and the means to help the children who are dying of starvation?

With rBGH in particular, some farmers have seen an increase in udder diseases, and others have noticed a decrease in the number of years that a cow can be viably milked. The increase in productivity is unquestionable, as are Monsanto's profits from its usage. In their 1997 report they state that "net sales in 1996 benefited from higher sales of Posilac bovine somatotropin,"[17] or rBGH. I believe that today in America, rBGH is not necessary, and also that before we encourage its use in other nations, where there may be benefits to it, we should demand further testing. We need to be assured that the potential long-term health hazards have been fully analyzed and that we understand what they are.

A SNAPSHOT OF MONSANTO: ONE OF THE HANDS THAT FEEDS US

Monsanto, with assets worth over $10 billion, net sales over $7.5 billion, and a net income of almost half a billion dollars, is one of the predominant companies in the biotechnology sector of the agriculture industry. Among its directors are the chairman of the Department of Genetics at Harvard Medical School, the CEO of Citicorp and Citibank, a former U.S. secretary of commerce, and the former administrator of the EPA. Together they make up a powerful scientific, political, and financial team, with its primary focus on controlling a large portion of the germplasm market. As such, it has sold off large segments of the corporation to raise cash to pay for seed company acquisitions. In November 1998 the *Wall Street Journal* reported that Monsanto would lay off between 700 and 1,000 workers, sell about $1 billion worth of businesses, and raise as much as $4 billion with equity and debt offerings, all in an effort to "monopolize" the seed industry.[18] In Monsanto's company report it defines itself as a "Life Sciences Company," because it is "engaged in three historically separate businesses—agriculture, food ingredients and pharmaceuticals—that now have begun to share common technologies and common goals. The goals are to help people around the world lead longer, healthier lives, at costs that they and their nations can afford, and without continued environmental degradation."[19]

This former chemical company, which has metamorphosed into a "Life

Sciences Company," could one day be controlling the majority of our food supplies. Are these the hands in which you want to put the lives of your family and friends?

THE FARMERS OF THE FUTURE:
NURTURING NATURE OR TEST TUBE TERRESTRIALS?

Monsanto's 1997 annual report states that "global planting of genetically improved crops increased from 5 million acres in 1996 to 30 million acres in 1997 and over 65 million acres in 1998. Crops with Monsanto traits account for approximately 50 million of those acres." The report goes on to say that

> [Monsanto has] scarcely begun to realize the potential of biotechnology and genomics. Monsanto's top two biotechnology product groups—Roundup Ready® products ... and Bollgard®, New Leaf® and Yield Gard® products, which resist insects—are each based on only one gene out of 40,000 in a plant. In 1997, we introduced cotton with both Bollgard® and Roundup Ready® traits, the first plants with specialized combinations of multiple traits. The ability to "stack" multiple genes in a single plant will allow Monsanto to commercialize more products faster and create far greater value for our customers.[20]

Marc Lappé and Britt Bailey report in *Against the Grain: Biotechnology and the Corporate Takeover of Food* that "according to Hartz Seed Company, a Monsanto subsidiary ... [by] the year 2000 Monsanto plans to have 100 percent of soybeans in the United States converted with the Roundup Ready gene technology."[21]

Traditionally, a farm has been a place where food grows according to the seasons and rhythms of nature. The struggle for life begins during the first stages of seed growth and continues until after the plant reaches maturity, reproduces, and finally dies. At any time in the growth cycle, a plant can be affected by a wide variety of weather conditions, diseases, and insect infestations. What appears on the surface to be the model of tranquility is in reality a system of ordered chaos. Companies that focus on bioengineering present a very different model. Martin Teitel illustrates it perfectly when he says,

> the business cycle ... involves acquisition, transformation, and sales. Business's bottom line is profit, and its production line is carefully controlled so that at any point in time the right number of widgets are moving through to ensure a steady profit. Predictability and uniformity characterize successful business systems. Attempts to adapt nature to the business cycle to make her products predictable, uniform, and consistently available, con-

tribute to the biodiversity crisis. When we pile melons and lettuce into our grocery cart in midwinter, or acorn squash and cabbage in midsummer, we encourage farmers and food retailers to override nature in deference to our gustatory desires. If we instead fill our kitchens with nature's seasonal provisions, we can steer the food industry toward greater respect for nature's cycle.[22]

But what do consumers want the farmers of the future to be growing, and at what price? Do we want fruit-year round that is grown from bioengineered seed so that it never freezes? Some of the new gene-spliced laboratory crops have been given attributes that would allow them to survive harsh weather conditions, as with tomatoes spliced with "antifreeze" genes found in flounder, or resist natural disease, as with potatoes altered to accept chicken genes for disease resistance. Is this our future? Do we want nature to dictate how we eat, as it has done for thousands of years, or do we want to engineer the system to allow us to have anything at any time? If we do that, the price we pay will be measured in levels of environmental destruction and in the health of our population. And yet, if we do not avail ourselves of emerging technology, there is a chance that the price we pay in human life will be unimaginable.

To visualize the farm of the future, imagine row upon row, hundreds of thousands of acres in area, of monocropped vegetables. Broccoli, tomatoes, and corn would be grown virtually year-round from the same seed stock, each identical to the one that came before. Surrounding the fields would be separator rows of nothingness. Between rows would be no sign of life because bioengineering will allow herbicide use that will kill all vegetation except for the predominant crop being grown. The ground will be extremely flat, and miles upon miles of it will be planted by laser and computer. Cows, pigs, and chickens will each be of the same breed and will all look the same across their species.[23] Pesticides and herbicides might be sprayed by satellite positioning systems. Farmers will have the potential for enormous production; perhaps robots will even replace farmworkers. A synthetic food supply may be on the horizon.

Of course we might envision an alternative model of the future, one that includes low-input, sustainable farming practices. Rodale Press published a report in 1997 that proposed an alternative:

Proponents of toxic pesticides and synthetic fertilizers claim that farmers could not grow enough food with organic methods to feed the people of the Earth, therefore, chemical use is necessary and justified. The longest-running experiment in the U.S. comparing chemical and organic farming is the Rodale Institute's Farming Systems Trial. For over 15 years, it has compared three farming systems side by side at the Rodale Institute Experimental Farm in Pennsylvania. One field has been farmed by conventional rotation of corn and soybeans, using chemical fertilizers and herbicides. Two other fields rotate corn, soybeans, grains, and hay, one using legumes for fertilizer, and the other manure, and both using mechanical weed controls. The organic fields produced the same yields as the chemical fields in normal years and greater yields in dry years. The soil quality of the organic fields was higher, holding more water and air, and suffering less erosion, enabling the organic fields to out-produce the chemical ones during years with low rainfall.[24]

Traditional agriculture utilizes over 80,000 species of plants, while industrial agriculture grows most of the food on our planet from just 150 varieties. As consumers we have the power to decide which we want for our future. I believe that it is very possible to combine the best of both worlds to create a sustainable, low-input agricultural model that would make the most of emerging technologies while maintaining strict control over the introduction of new, potentially harmful practices. However, we cannot do so without a strong commitment from government, consumer, and corporate groups to work together toward a positive end.

WHAT'S THE BEEF WITH BEEF, POULTRY, PORK, AND LAMB?

Max Finberg, special assistant to the director of the Congressional Hunger Center, told me he stopped eating beef because it takes as much as seven pounds of grain to produce a single pound of meat. He would rather see the grain put to better use feeding a larger number of hungry people than one pound of beef ever could. Without a doubt, there are more efficient ways to produce food. In this country, however, beef is a mainstay. We eat approximately seventy-five pounds of beef per person per year. Using the above figures to measure the environmental cost of per capita beef consumption, each person in the United States consumes 525 pounds of grain and over 13,000 gallons of water—just for hamburgers, steaks, stews, and roasts! Those cows excrete 230,000 pounds of manure per second,[25] and to satisfy our addiction to beef, 100,000 cattle are slaughtered every twenty-four hours. Livestock production consumes almost half the energy used in all U.S. agriculture and 70 percent of the country's grain. The grain, soy, and other feeds consumed by our livestock approach 200 million tons a year, which represents about one-half of all the harvested acreage in the country. The United States alone uses 2.9 billion bushels of corn to feed 39.6 million calves and 22 million cattle. Planting, watering, fertilizing, spraying with pesticides, and harvesting this vast area of cropland has enormous environmental implications. Chemical companies also benefit greatly from having land farmed to feed animals, since animal feed carries far less stringent pesticide tolerances than does grain intended for human consumption. The fossil fuel used to grow one pound of feedlot-fed beef would grow about forty pounds of soybeans. In fact, according to a joint Departments of the Interior and Commerce report, "One-third of the value of all raw materials consumed for all purposes in the United States is consumed in livestock foods."[26] To compound this problem, livestock also needs a lot of space; an acre of land can be used to grow 20,000 pounds of potatoes, or 165 pounds of beef.

The land and the environment are important considerations, but so too are the chemicals, hormones, feed additives, and pesticides cattle ingest. Many people believe that a significant portion of these substances eventually ends up in our food supply. Some of the items that may be fed to the steer from which your hamburger or steak is produced include bloodmeal, or dried blood from animal-processing plants; feather meal, or hydrolyzed poultry feathers; and meat and bone meal, which consists of rendered and dried product from animal tissue, dehydrated cattle manure, animal fat, poultry fat, and poultry manure. All of those ingredients might have been fed to the steer that was processed into your child's hamburger or the cow that gave the milk in your refrigerator.

Consumers who, unlike Max Fineberg, do not want to give up eating beef are seeking alternatives. Mel Coleman, founder of Coleman's Natural Beef, has dedicated the past two decades to producing natural, sustainable beef in a profitable fashion. In Stephen Voynick's book *Riding the Higher Range: The Story of Colorado's Coleman Ranch and Coleman Natural Beef*, Mel explains his philosophy: "What we are doing is going back to the way we raised cattle in the 1930s. After World War II, the availability of new drugs and chemicals coincided with a need to produce huge quantities of cheap food. American agriculture used chemicals to meet that challenge, but it paid a price. We have lowered the quality of our food, degraded our environment, and lost many of our traditional values concerning land, rural life and the treatment of animals. The saddest part is that we never needed all those chemicals to produce good beef."[27]

With the help of his wife and family, Coleman has brought "natural" beef onto the commercial market, and they do it sustainably. Voynick explains, "More than 350 Coleman Certified Ranchers now raise nearly a quarter-million cattle without growth promoting hormones or antibiotics. And they employ modern range management methods to improve millions of acres of range. All receive above-market cattle prices, making them more likely to keep their lands productive. Each certified rancher is helping to return the emphasis in cattle production from grain feeding back to the use of a natural, renewable resource—grass."[28] This certainly seems like a sustainable system, but is it productive enough to feed our growing population?

Like Mel Coleman, agribusiness is trying to build a better cow (or pig or chicken, or sheep—who can forget Molly and Dolly, the sheep cloned in the 1990s?). "Researchers are developing genetically engineered 'super animals' with enhanced characteristics for food production," reports Jeremy

Rifkin in *The Biotech Century: Harnessing the Gene and Remaking the World*. He goes on to say,

> At the University of Adelaide in Australia, scientists have developed a novel breed of genetically engineered pigs that are 30 percent more efficient and brought to market seven weeks earlier than normal pigs. The Australian Commonwealth Scientific and Industrial Organization has produced genetically engineered sheep that grow 30 percent faster than normal ones and are currently transplanting genes into sheep to make their wool grow faster. At the University of Wisconsin, scientists genetically altered brooding hens to increase their productivity. As brooding hens make up nearly 20 percent of an average flock, researchers were anxious to curtail the "brooding instinct" because "broodiness disrupts production and costs producers a lot of money." The new breed of genetically engineered hens no longer exhibits the mothering instinct. They do, however, produce more eggs.[29]

It seems that we are able to remake most mammals to be and do anything we desire, from growing their hair faster to laying more eggs with less fanfare. From producing more milk to growing more edible meat, no changes to nature and other mammals seem beyond our capacity to control.

Thanksgiving has its roots in the beginning of our agricultural history. After the Plymouth colonists reaped their first harvest in 1621, Governor William Bradford proclaimed a day of thanksgiving. The Pilgrims entertained ninety-two Native American guests, including Chief Massasoit, at a Thanksgiving "dinner-breakfast," which included wild turkeys shot by the colonists. By most accounts those early birds looked nothing like our turkeys; they are reported to have resembled large multicolored chickens. In *Enduring Seeds*, Gary Nabhan describes the mutation of the turkey from wild bird to modern dining fare: "Today's turkeys can grow to 22 pounds in as many weeks, and have become completely standardized in look and taste."[30] The beginning of the drive to create uniformity within the species dates back to the mid-1800s, when poultry exhibitions became popular; the American Poultry Association manual went so far as to mandate their "Standard of Perfection" for this once regal bird.

Modern-day turkeys are believed to be among the stupidest of barnyard creatures. A National Research Council publication reports that "they often have to be taught to eat and drink and they become easily lost." Another illustration of the bird's intelligence: "Young chickens are put into the pens so the turkey poults will learn to peck by watching the chickens. Without chickens, they stand in piles of grain and die of starvation. Then whenever it threatens to rain, we rush out into the fields to herd the turkeys back into the barn. It is only after the slow realization that it was water, not food, moving on the ground, that the entire flock in incredibly orchestrated movement raise their necks from the ground, tilt their heads to the sky, and open their beaks wide to the falling rain until they drown."[31]

In the mid-1900s New Mexico ornithologist J. S. Lignon wrote, "Probably no other bird or mammal excels the turkey in alertness"—an amazing assertion, given the story just told. "It can instantly detect the slightest movement of an object in the scope of vision. Even the shriek of a chipmunk, chatter of a chickadee, or the scolding of tiny bushtits is sufficient to cause all turkeys within hearing to snap to attention. If the danger proves real, the skill and speed with which all disappear is astonishing."[32] It does not seem possible that these stories relate to the same animal, and yet the difference is solely the result of intensive genetic tinkering and cross-breeding. Most of the 124 million turkeys raised in the United States today belong to a handful of uniform breeds, all of which have been bred for our dining pleasure and ease of production. Possibly foreshadowing the farms of the future, turkeys today can hardly stand up or walk because their breasts are overtly disproportionate to their body mass. The broad-breasted white turkey has been so overdeveloped for meat production that it is unable to mate naturally because of its shape. Only through artificial insemination can it reproduce itself.

Another wild bird that has been domesticated and raised for food is the chicken, of which Americans consume 49 pounds per capita annually. I interviewed Richard Roop, director of corporate quality assurance for Tyson Foods, one of the largest chicken producers in the United States. Tyson's 1997 sales were approximately $8 billion. The company processed over 2.1 billion chickens, 8.9 million turkeys, and 1.6 million hogs. With over eighty processing plants and almost forty feed mills, this agribusiness giant controls over 20 percent of the chicken market and produces an incredible volume of chicken manure. Rick and two of his coworkers were kind enough to spend over an hour with me on the phone, giving me a crash course in Chicken 101. During our interview I learned that approximately 20,000 to 40,000 chickens live in a "chicken house," which can be likened to a metal barn, 40 by 450 feet in size, with large barrierlike cages to keep the chickens from flying out. One chicken house with 20,000 chickens produces approximately 100 tons of manure a year, which Tyson representatives told me is later utilized as fertilizer.

For the birds to grow to optimum size, approximately 3.8 to 4 pounds liveweight, takes roughly 38 to 42 days, during which time they will eat about 7 pounds of feed each. The feed consists of numerous forms of protein as well as canola, cottonseed, poultry by-products, and feather meal along with synthetic amino acids. Fats and oils are also used as feed supplements and may be composed of restaurant grease or other animal/vegetable sources. Rick and his coworkers explained to me that almost all of the parent stock of their flocks come from one of five companies, and that genetic selection is the reason the poultry industry is so competitive. Every year it takes Tyson one day fewer to raise a chicken to the same weight, and typically, for the past twenty years, it has reduced the feed conversion rate by one to two percentage points annually. Yet even as the chicken industry becomes more efficient, the volume of resources used may seem startling to some. For instance, the typical chicken house, with its 20,000 chickens and 100 tons of manure, may produce just under 200,000 chickens in a year. Those chickens will consume approximately 1.4 million pounds of feed and drink 1.6 million gallons of water during that year. This is only representative of one chicken house of the forty owned by Tyson. Extrapolated, that would mean that the entire chicken industry uses 254 million tons of feed and 290 million gallons of water each year. Chicken manure produced during that time would reach nearly 20,000 tons.[33]

Of course, chickens have not escaped the same kind of cross-breeding that has made turkeys the stupidest of barnyard animals. In fact, chickens are the subject of a USDA research project that will eventually map out their complete genetic structure, making it easier to manipulate their characteristics, creating bigger, leaner, healthier chickens that reach maturity with greater rapidity. Some of this could be seen as beneficial, to the chicken industry as well as the world's growing population; but it is easy to see this kind of scientific development being utilized simply for profit, without regard to the health of the animals or the people who consume them. A recent article in *Time* magazine reports that research by scientists at the Salk Institute (whose focus is unmistakably on human health) has led to the discovery of the genes that dictate the growth of chicken legs. They successfully created embryos that grew four legs instead of two.[34] While this research was done only for the benefit of human health, one wonders how long it will be before these four-legged chickens are all the rage in supermarkets around the world. How long will it be before children do not know that chickens ever had wings at all?

MANURE LAGOONS FROM A BAY OF PIGS

The pork industry, which creates net sales of approximately $30 billion, is a relatively small part of agribusiness, but one which has undergone double-digit growth and tremendous change over the past several years. Currently, the time from birth to slaughter can be as short as five and a half months. Nevertheless, roughly 1.3 billion tons of pork manure was produced last year—5 tons of manure for every American. The manure is transported to and held in "lagoons" that are typically 25 feet deep and can be as long as a football field. Just like any other manmade pools, they have the potential to spring leaks from time to time. The *Boston Globe* reported one such instance: "In 1995 in North Carolina, a record hog waste spill of 35 million gallons killed 10 million fish and closed 364,000 acres of coastal wetlands to shellfish harvesting. Spills in Iowa, Minnesota, and Missouri in 1996 killed more than half a million fish. In Virginia, Maryland, and North Carolina, waste from hog and chicken factories is widely believed to have caused pfiesteria, which has killed more than a billion fish and excess nutrients flowing down the Mississippi River are believed to have resulted in a dead zone the size of New Jersey in the Gulf of Mexico."[35]

These manure lagoons are the direst potential environmental threat posed by factory hog farms. In the five to six months of a hog's life, it produces two to four times as much waste as a human. In the 1997 *Newsletter of the Pennsylvania Association for Sustainable Agriculture (PASA)*, Brian DeVore wrote, "The manure sits and festers in an anaerobic soup for several months before being pumped out onto farm fields where it becomes the basis for growing crops and 'replenishing' soil. The larger lagoons hold tens of millions of gallons, cooking up a stench so powerful that even veteran hog farmers complain about it. These hog lagoons emit at least 130 different gases, including ammonia, yet state and federal clean air rules generally fail to address them, in fact, factory farm supporters have acknowledged that factory hog farms pose an environmental risk, but say the financial payback makes it worthwhile."[36] Indeed, this profit is why in 1998 over 300,000 new animals went into production, and sow numbers for the largest producers grew by 19 percent. When we take into account hog farm subsidies and price supports (see p. 150), it is hard to fathom the logic of this method of agriculture.

THE OLD BAIT AND EXCAVATE

The ocean suffers from the same devastation as the land—chemical pollution, overmining and overharvesting of resources, and unsustainable farming practices all plague the oceans as well. "Overfishing is a growing global problem," said Mark Kurlansky in *Food and Wine* magazine:

> About 60 percent of the fish types tracked by the Food and Agriculture Organization (FAO) of the United Nations are categorized as fully exploited, over-exploited, or depleted. The U.S. Atlantic coast has witnessed a dramatic decline in its bluefin tuna population. Mid-Atlantic swordfish stocks are diminishing. And red snapper, which gets caught in shrimp nets, is in danger of commercial extinction in the Gulf of Mexico. We spent the Eighties devouring blackened redfish until the redfish population of the Gulf of Mexico was almost gone. Orange roughy was introduced in the Seventies and immediately gained such popularity that five tons an hour were being taken from the ocean depths near New Zealand; in 1995 the catch nearly vanished.[37]

For thousands of years fish and shellfish have been a naturally replenished mainstay in the diets of the world's people. Not so now.

Steve Connolly owns one of the largest seafood companies in New England. He ships fish all over the United States, as well as to other countries, and has been in the seafood business in Boston since the early 1950s. Although he has been in the market for less than fifty years, he has seen tremendous change in the industry—from availability of product to wild price fluctuations. He told me that in January 1955, they "landed 358,000 pounds of large cod; 277,000 pounds of market cod; 2,795,000 pounds of haddock; 2,248,000 pounds of haddock scrub; hake: 184,000; pollock: 1,724,000; and redfish: 159,000—all in 31 days in January!" In January 1997 the entire state of Massachusetts (not just Connolly's business) brought in 1,037,786 pounds of cod; 119,231 pounds of haddock; 84,000 pounds of hake; 300,240 pounds of pollock; and 28,591 pounds of redfish.[38] The combined cod and haddock catch, often thought of as the mainstay of that fishery, is down almost 80 percent in less than forty-five years.

I find that almost impossible to fathom.

As a result, prices have increased by thousands of percents. For example, lobster rose from 65¢ to $7.35 per pound, cod went from 10¢ to $5.82 per pound, and yellowfin tuna rose from a mere 29¢ to $8.97.

Over the years, technology has advanced ahead of the ocean's ability to replenish itself. The competition to land the biggest and greatest number of fish has given rise to larger and more sophisticated fishing vessels, to satellites, and to military-style sonar and "spotter planes" to track schools of fish. Factory ships, like factory farms, exploit resources. They are able to put out to sea for months at a time, and between "bottom drag" nets and "quick freeze" can actually harvest as much as 500,000 pounds of fish in just one haul. Every year factory ships harvest more than 1.7 billion pounds of fish from U.S. waters. Once an industry that focused primarily on local fish stocks, fishing has metamorphosed into a global system of intensive agribusiness. In this expanded system, worldwide supply and demand has great influence on price fluctuations in the marketplace. For instance, a single bluefin tuna caught off the coast of New England recently sold for $83,500, nearly $117 per pound when flown to Japan. The 715-pound giant was to be reduced to 2,400 servings of sushi, which, because of the exceptional quality of this individual fish, would be served to elite businessmen and government officials for $75 per serving, bringing in altogether an estimated $180,000.[39] Competition among fishermen is fierce, and the stakes are higher than ever. Adding this global element to an industry only intensifies the crisis we are seeing in our oceans and rivers. Each year, in a race to hoist the largest catch, fishing boats draw up an estimated twenty-seven million metric tons of marine life that, dying or dead, is thrown overboard because it does not meet the standards set by fishing regulations, quotas, and licenses. That is one-quarter of the entire global catch. When fish are removed from the ocean, without regard to their age or level of maturity, the result is a diminished wild population that is unable to replenish itself.

The United Nations–sanctioned "International Year of the Ocean" brought one fish in particular into the limelight. In 1998 a number of American chefs signed on to a "Give Swordfish a Break Campaign," whose goal it was to give the North Atlantic swordfish population an opportunity to rejuvenate itself. Up until the 1950s, the majority of swordfish were caught by harpoon as the mature fish came up to the surface to sun themselves. Smaller fish, which are much less inclined to engage in this behavior, were able to stay out of harm's way. As the technology advanced and long-line fishing became the norm, swordfish were caught in greater numbers and smaller sizes. Where once the harpoon fisherman actually saw, chased, and killed his prey, long-lines, often miles long with hundreds of hooks, made the hunt for swordfish obsolete. The fisherman wielding a harpoon could make judgments about the fish he hunted; but anonymously placed hooks do not discriminate. Swordfish caught in the North Atlantic in the 1960s weighed more than 250 pounds; due primarily to overfishing, the average swordfish caught in the North Atlantic today weighs 90 pounds. Nearly two-thirds of the swordfish caught today are too young to breed, and because of long-line fishing methods U.S. fishermen discarded 40,000 North Atlantic swordfish in 1996, principally because they were too small legally to bring to market.[40]

The grouper, a less well known fish, is in danger because it is tragically misunderstood. Doug Raider, senior scientist of the North Carolina Environmental Defense Fund, who has spent a great deal of time studying the oceans' fish populations, explained the plight of the grouper to me: it is "a fish that does some really cool things ... they are protogenous hermaphrodites. What that means is that when they become sexually mature they are females, and only the biggest, oldest individuals ever become male. They used to be called 'supermales,' and ... would manage a harem of females. However, new size limits ... require fish to be larger to be taken. This has resulted in the older, male fish being harvested. That, combined with the fact that groupers aggregate to mate, has made some species endangered. The net effect has been that the sex ratio has plummeted; where one in five fish became male, now the ratio is one in twenty or less."[41]

The grouper and the swordfish are certainly good examples of how even our best intentions are not necessarily best for a particular species. Each fish requires its own set of research objectives. The grouper might be saved if, instead of taking larger fish, as has become the norm, we would take smaller fish in order to save the males. Likewise, the idea behind the "Give Swordfish a Break" campaign is a good one, but there are some inherent problems there too. To begin with, Pacific swordfish are not endangered, and there should be no reason to stop buying them provided they meet the standards set out by the fishing industry. Because the swordfish campaign was only active in the United States, there was no incentive for foreign fishermen to stop catching small swordfish. By imposing this sort of ban on fishing of any kind, we not only hurt our fishermen but do not necessarily do anything to save the fish. True sustainability means creating a balance. In this case, better fisheries management on a global scale is what is needed. At this point, it seems unclear what—other than hurting some of our own fishermen and bringing media attention to the fish—the swordfish campaign actually did.

There is no doubt that we are overfishing the oceans, but the true absurdity is that we are actually paying with tax dollars to have that accomplished: "A 1989 study by the United Nations Food and Agriculture Organization estimated that it cost about $92 billion to operate the global fishing fleet. Revenue, on the other hand, was only $70 billion; much of the difference was made up by subsidies from governments to fishermen and boat builders. Throughout the 1990s this type of disparity has held true."[42]

Today there is little guesswork involved in fishing. As a response to overfishing, numerous rules, regulations, and quotas have been implemented. Though fishermen are largely skeptical of the scientific methods of fisheries management that are instrumental in setting quotas, and which could be as much as 100 to 200 percent erroneous, most abide by the regulations. Of course this does not help their bottom line, and many have been forced to stop fishing altogether because some of the quotas make it impossible for them to earn a decent living. All of these factors combined have seen the world fish catch level off at 100 million tons per year. Farmed fish now makes up 30 percent of that amount, a figure that has doubled in the past decade. As our natural resources become diminished, we are becoming more dependent on aquaculture, which many believe to be quite unsustainable.

For the better part of the year in most of the continental United States, the majority of fresh salmon consumed is raised on a farm. This was not always the case. Salmon is a saltwater fish that spends its youth (approximately two years) in freshwater rivers and streams where it reaches maturity, and finally swims downstream to the sea. A full-grown fish will remain at sea for several years, but returns to the exact location of its birth to spawn. For Pacific salmon, spawning marks the end of life; Atlantic salmon, however, often spawn and return to the open ocean. The building of dams has severely limited the fish's ability to return to its spawning grounds, and the population has dwindled. Over the last sixty years in Maine, the catch has decreased from 150,000 pounds to less than 1,000. The North Atlantic Salmon Fund has surmised that in the past twenty years, the total catch of true Atlantic salmon (excluding fish escaped from fish farms) has fallen by 80 percent, from more than 4 million to 800,000.[43] The same dramatic loss has occurred on the West Coast. In response, the fishing industry has constructed salmon ladders to help the fish circumnavigate dams and return to their breeding grounds, but the population has never recovered. Wild salmon have been pushed to the brink of extinction in many areas, and in 1999 nine species of Pacific salmon were added to the endangered species list.

Salmon farms are generally comprised of a series of heavy-gauge netpens, which, when anchored in coastal waters, can hold up to several thousand fish at one time. I have been told that to best understand the fishes' living conditions on a salmon farm, I would have to get into an average-sized bathtub along with two twelve-pound fish. Very tight living quarters, to say the least. Though salmon farming might initially appear to be a great solution to our sudden lack of wild fish, it may not be. Like land-based farming, aquaculture can produce a great deal of unwanted pollution, caused mainly by insufficient water circulation. Water polluted with unconsumed feed and excrement becomes high in nitrogen, which ends up in open waterways surrounding the netpens. Another concern in aquaculture, particularly of carnivorous fish like salmon, is that our use of resources is wasteful. For example, to produce one pound of salmon takes two and a half pounds of fish out of the wild, which is made into fish meal and oil and fed to the salmon. As Lester Brown points out in *State of the World 1998*, "Once we put fish in ponds or cages, they have to be fed. Fishponds are, in effect, marine feedlots, competing with humans and producers of poultry, pork, and beef for grain."[44]

The fish used in the production of fish meal and fish oil are "small pelagic fishes" like anchovies, sardines, jack mackerel, and herring. Fish meal is used to nourish not only other fish but also pigs, chickens, and to a lesser extent cattle. Some argue that because the species utilized for fish meal are not in great commercial demand, we ought not to worry about the way they are consumed. In the aquaculture industry in particular, utilization of small feed fishes accounts for only 15 percent of the 31 million metric tons of fish harvested annually for animal feed; but that same figure accounts for nearly a quarter of all the wild fish harvested annually.

Shrimp aquaculture, often practiced in emerging nations, has been widely criticized—although it is one of the most efficient uses of protein in animal agriculture—for perceived environmental damage. It first became profitable in the 1970s and has since mushroomed into a widespread tropical enterprise. Thailand, Indonesia, China, India, and other Asian nations are now home to over 3 million acres of shrimp ponds. Nearly half a million acres of coastline in the Western Hemisphere have been similarly transformed. Like other forms of aquaculture, shrimp farming is considered extremely detrimental to the coastal waterways it inhabits. Mangrove forests, in particular—which actually work to protect coastal waters by acting as natural filtration systems—have been destroyed to make way for shrimp ponds. Traditionally, ponds rendered useless due to pollution and disease were simply abandoned, and the mangroves were not replanted. It has also been standard practice for shrimp farmers to disgorge into the oceans between 10 and 30 percent of the water in their ponds every day. The pollutants—feed, feces, and ammonia and carbon dioxide by-products—released so frequently caused great damage to coastal waterways. Recently, however, some Asian countries have made progress in teaching local shrimp farmers the importance of the mangroves, and most shrimp farmers are only disgorging between 2 and 5 percent of their pond water daily. But environmentalists remain concerned. One reason is reports stating that as much as 30 percent of the feed used in shrimp farming is never consumed.

David Wills, consultant to Darden Restaurants Inc., spends much of his time touring and assessing worldwide shrimp aquaculture facilities. His research has led him to believe that corporate aquaculture has actually been beneficial to the people and economies of emerging nations. The multinational corporations that own shrimp farms are building roads, schools, hospitals, and housing, Wills says; "They are bringing civilization to coastal areas that frankly did not have it before." Wills also defends shrimp farmers, observing that the majority of mangrove destruction cannot currently be laid at their door. Most of the trees are actually cut for firewood by the native populations for cooking fuel, and are "burned and gone the next day."[45] He also told me that most of the shrimp farms in question have begun reforestation efforts to increase the number of trees at water's edge. Claude E. Boyd and Jason W. Clay agreed with Wills in the June 1998 issue of *Scientific American*, noting that "shrimp farming alone appears to be responsible for less than 10 percent of the global loss [of mangroves]."[46] Not everyone in the food service industry agrees with Wills that the shrimp companies are enhancing the life of the native population. In fact, at the 1998 Chefs Collaborative 2000 retreat, some expressed their feeling that forcing our way of life onto native populations could actually be quite irresponsible. I believe it is naive, however, to think that other countries would not be looking to improve their global communications and commerce systems, even if there were no shrimp farming.

The word on aquaculture is not all negative. Trout has long been farmed in America's rivers and ponds, and recently hybrid striped bass has become a part of the landscape of aquaculture. Jim Carlberg is the president of Kent Seafarms, a freshwater striped bass farming company in southern California. Carlberg and his colleagues have been working with the Department of Commerce and the USDA to develop innovative, low-cost approaches to wastewater treatment that are compatible with agricultural application, he told me: "We not only recycle and reuse some water, reducing our amount of makeup water [water required to make up for water pumped out for cleaning purposes], but also make our effluent water available for irrigation. . . . By developing a fairly elaborate network of pumping and piping systems, we have been able to channel our water and deliver it to adjacent farmers, and they are now applying that directly to their crops, as a supplement to fertilizer."[47] Jim believes that such an integration of aquacultural systems into land-based farming could go a long way toward creating an entirely sustainable agricultural system. Currently, a stumbling block to Kent Seafarms' sustainability is the fact that like salmon, striped bass are carnivores, and cause a similar net loss of wild seafood.

Michael Romano, chef at Union Square Café in New York City, believes that although the fishing industry is becoming more proactive in protecting wild stocks, aquaculture is here to stay: "There are always two sides to a coin. I know that I do not want the wild fish supplies to disappear, and it is so odd for us to have to talk about 'wild' fish, which is all we have known for most of our lives! Now you see it appearing on menus as if the term 'wild' meant something special, instead of the norm, which is really a tragedy."[48]

At this time there is no question that work is being done to help rejuvenate the wild fish population—from imposing catch quotas to outright bans on fishing certain species. But in case we do not have the capacity to increase and protect our wild fishes, bioengineering experiments may help produce hardier farm-raised fish to offset the decline. Jeremy Rifkin reports that fish have recently been genetically engineered to increase food conversion efficiency, tolerate increased cold and salinity, and become more disease-resistant. Biotechnicians have spliced an "antifreeze" gene from flounder into bass and trout so the fish will survive more frigid temperatures. Another team "inserted a mammalian growth hormone gene into fertilized fish eggs, producing faster-growing and heavier fish. Still other researchers are experimenting with the creation of sterile salmon [that] will not have the suicidal urge to spawn, but rather remain in the open sea to be commercially harvested."[49] Most of the genetically modified fish are being designed to live in commercial tanks and on fish farms. This may seem like a step in the right direction for the technology, the fish, and our diminishing resources, but this kind of experimentation has many potential downsides. The most frequently noted concern is that accidental escapes by transgenic fish into open water could cause native fish species to be outcompeted for natural resources. Left unchecked, the results of this kind of experimentation could be dangerous; given its potential for severe damage to wild stocks, it should be closely monitored and tightly controlled.

BORROWING THE FUTURE

The primary issue surrounding agribusiness and biotechnology is what the future impact will be. Are we investing in the future and revitalizing our resources, or are we "borrowing" from our children's children? We do not have the answers to these questions yet, but we can make some predictions based on past and present applications of agribusiness technologies. Since the majority of us are already consuming bioengineered products (Crisco, Kraft salad dressings, Nestle chocolate, Green Giant harvest burgers, Parkay margarine, Isomil and ProSobee infant formulas, Wesson vegetable oils, McDonald's french fries, Fritos, Doritos, Tostitos, and Ruffles chips all contain some form of genetically modified soybeans) as well as foods sprayed with herbicides and pesticides, we should all be taking a good hard look at what it means for the future of our resources. Nothing is ever as simple as it appears to be, and as we learned from our use of DDT, the damage to our planet—and ultimately our health—may not be immediately evident. What may appear to be a miracle could actually just be a hazardous experiment.

Bacillus thuringiensis (Bt), discovered in common soil bacteria in 1911, has been available in commercial formulations for insect control since the 1930s. The pesticidal properties of this naturally occurring substance seem almost miraculous, especially for IPM and other low-input farming. Recently, however, Monsanto has begun splicing the Bt gene into corn, cotton, and potatoes, giving the plants themselves the ability to act as their own pesticides. This may seem like a perfect application of our best technology, but it also raises some serious concerns. One major issue is resistance, which happens much faster when monocropped acreage is covered by Bt vegetables, as opposed to small-scale farmers utilizing Bt as part of an IPM plan. When the insects that eat these plants become resistant to the Bt gene's toxin, which they ultimately do, one very effective means of alternative pest control slowly becomes obsolete. Monsanto former vice president of marketing Phil Angel says of the risk involved, "Do we run the risk of building up resistance to the Bt gene that we use in our corn? The answer is, it is absolutely a potential problem. The use of these technologies requires the development of ever more toxic or powerful chemicals, because you are in fact having to overcome the essential resistance that builds up over time."[50] In 1997 a team of scientists reported that 1.5 out of 1,000 moth larvae carry the gene to overcome Bt toxin. That may not sound like much, but it is a thousand times more than expected, and it may be attributed to the use of Bt seeds. Some scientists estimate that the effective shelf life of Bt technologies is no more than ten years, because they expect that most insects will have developed a resistance by that time. The more crops produced with the Bt gene, the more quickly its capabilities are likely to become compromised. Monsanto suggests to its growers that they maintain "safe" zones of non-Bt crops as an area for insects to breed, but thus far, there is no evidence that this method is in any way effective, or that farmers are actually following the company's mandate. If ten years is the total estimated length of effectiveness for this technology under the best of circumstances, perhaps we are being shortsighted by allowing it to move forward at all. Is the short-term profit by companies like Monsanto worth sacrificing the long-term effectiveness of a product that is currently extremely helpful to the industry? What are the consequences of applying a technology like this one?

Because pesticides like Bt have a limited shelf life, many companies working in transgenics are traveling the world in search of indigenous plants that have beneficial agricultural properties. Jeremy Rifkin describes a recent occurrence in India:

The *neem* tree is an Indian symbol and enjoys an almost mystical status in that country, ancient Indian texts refer to it as "the blessed tree" and the tree that "cures all ailments." The tree has been used for centuries as a source for medicines and millions of Indians use it to cure acne and to treat a range of illnesses from infections to diabetes. The tree has proven particularly valuable as a natural pesticide and has been used by villagers to protect against crop pests for hundreds of years. In fact, this natural pesticide is more potent than many conventional industrial insecticides. The *neem* extract wards off more than two hundred common insects that harm crops, including locusts, nematodes, boll weevils, beetles, and hoppers. A number of studies have found the *neem* extract to be at least as effective as synthetic insecticides like Malathion, DDT, and dieldrin, but without the deleterious impacts on the environment.[51]

W. R. Grace, a company made infamous by the 1999 release of the film *A Civil Action* (in which W. R. Grace was found liable for chemical pollution that caused high rates of leukemia in a small Massachusetts town) has recently become interested in the neem tree. The company isolated the most potent ingredient in the neem seed, and then sought and received a number of process patents for the production of neem extract. This sets a serious precedent that could cause irreparable damage to the landscape, environment, and well-being of countries like India. Not only that, native peoples who have relied on plants like the neem tree may one day find themselves on the wrong side of the law.

Plants (such as Roundup Ready products) bioengineered with herbicide resistance may produce problems much like those of their pesticide-engineered cousins. In this case, instead of stronger bugs, there is fear of out-crossing, which creates "superweeds." Again, Angel explains that "The issue of out-crossing, whether in fact you can create some kind of super-weed, is a potential. A plant that is resistant to Roundup, for instance, could transfer that resistance to a plant in the same family that is not a crop plant, and it may be resistant to Roundup in a weedy area. If there is a situation in which a particular weed needs to be controlled, and it happens to be Roundup-resistant, then you would just use another herbicide to kill it."[52] If one chemical does not work, add another one to the mix—that ought to take care of it. Until, of course, the weeds develop resistance to that chemical—and the next, and the next. Studies have already shown that plants such as rapeseed that are engineered to be herbicide resistant may be able to pass that resistance along to their cousins and produce "superweeds," resistant to high doses of herbicides. Another potential problem is that farmers may be more likely to spray larger quantities of chemicals on their herbicide-tolerant crops because they do not risk losing their crop plants due to overapplication of herbicides. After all, more chemicals kill more and stronger weeds. Chemical companies claim that increased use of herbicide-resistant plants will lead to less, not more, use of their products, but why would they want to sell less product and make less money? Jeremy Rifkin reports that Monsanto has already begun dealing with weed tolerance. The company "has applied to the regulatory authorities in a number of countries, requesting an increase in the residue limit for its Roundup chemical on crops."[53] This will allow farmers to use more chemicals to kill weeds, and it will not hurt Monsanto's bottom line either.

Of course, once we apply all these herbicides and pesticides to our crops, no matter what arguable good they may be doing for farmers and our food supply, they are wreaking havoc on our water supplies. Lester Brown reports that "agricultural runoff in the Mississippi River basin is now so extensive that when the river enters the Gulf of Mexico, the overfertilized brew of nutrients it carries sparks huge alga blooms, which deplete the water of oxygen and create a 'dead zone' of some 17,600 square kilometers, nearly the size of New Jersey."[54] If biotechnology could in fact reduce the amount of chemicals we use, it would be, as Phil Angel says, "an important part of how the world meets its food needs in the future. It is a powerful technology which can deliver enormous economic benefits to farmers, and it can have a powerful impact on people's lives in a positive way."[55] Yet in our headlong rush toward the future, we seem to be throwing caution to the wind. According to Hendrik Verfaillie, senior vice president and chief financial officer of Monsanto, "The biggest mistake that anyone can make is moving slowly, because the game is going to be over before you start."[56] His words echo in my mind.

Is this a game? A race? It seems to be, and it is speeding its way around the globe on trucks, trains, and planes. Fossil fuels are not a renewable resource, and they are extremely undervalued. Transported and imported products, because they have been grown by agribusinesses in great quantities, are often priced lower than local products grown on smaller farms. We burn fuel to get strawberries from California to the East Coast because they are cheaper. How can that be? Fuel is inexpensive in the short term, when we ignore the cost of environmental damage or the notion that we may one day run out of oil and gas. If we had to pay the cost of environmental remediation caused by transporting food long distances, then food with frequent flyer miles would be more expensive than food grown around the corner or down the street.

As biotechnology expands and develops, thoughtful food choices should be even more vital than they have been in the past. Perhaps the question is not whether we can survive without biotechnology, but which types will be the most beneficial to our future. Jeremy Rifkin asks, "Will we use our new insights into the workings of plant and animal genomes to create genetically engineered 'super crops' and transgenic animals, or new techniques for advancing ecological agriculture and more humane animal husbandry practices? Will we use the information we are collecting on the human genome to alter our genetic makeup or pursue new sophisticated health prevention practices?"[57] To be sure, we should not work to destroy an industry that may hold the answer to reducing the amount of chemicals introduced into the environment annually, nor should we deny emerging nations lacking an adequate food supply a chance at alleviating hunger or malnutrition among their populations. The future of biotechnology in this country with regard to the food supply and food service industry lies in the products we allow onto the market. As consumers, we must vote with our food dollars and demand that biotechnology give the products and results that are best for the long-term health of the planet and all those who inhabit it.

BIOCOLONIALISM

Biocolonialism is a set of policies by which a nation maintains and controls the living plants, food crops, and industries of another country as a means of supplying the demands of its population. Coffee, fish, shrimp, bananas, strawberries, and oranges are but a few of the items grown to order for the United States and imported from other countries. Kevin Knox, vice president and coffee buyer for Allegro Coffee, spoke to me at length about this phenomenon:

> In most countries, coffee is a colonial crop, it is a cash crop that was imposed on a country and its people. This is a really important thing for people to understand, because in every country in the world other than Ethiopia, the people who grow the coffee do not drink it. They do not know anything about it, other than [having] a profound understanding of the horticultural aspects. They know about the diseases that affect it, and what the appearance should be like, but they have never tasted their own coffee. The nature of the crop is as a cash crop for export, the high-grades have gone and continue to go to the First World, where they fetch high prices. The dregs, the leftovers, are what are kept for domestic consumption.[58]

Kevin went on to tell me that coffee served on a plantation in Guatemala or Antigua will more than likely be Nescafé.

Eighty-five percent of the world's fish is imported and exported globally. Japan maintained its 154-pound-per-capita intake by importing $18 billion of fresh fish, shrimp, and surimi, among other fish products, in 1995.[59] Japan is by far the largest consumer of fish worldwide (consuming upward of six times the global average), but other industrial nations are increasing demand. Now, reports Lester Brown, developing countries are solid net exporters. Chile, China, and Thailand are responsible for the lion's share of the net trade balance, and their profits come from a fairly narrow product range. Thailand's rapid growth in the past decade, for example, has been achieved through the development of a canned tuna industry and a 1,000-fold increase in giant tiger prawn. In Bangladesh, China, India, Indonesia, and Thailand, farmed shrimp are one of the leading food exports. Developing countries trade more than 65 percent of the world's fresh and frozen tuna, 85 percent of the canned tuna, and 84 percent of the world's frozen shrimp and prawns.

Brown goes on to explain, seconding Darden Foods consultant David Wills, that this type of agricultural development is a double-edged sword for developing countries. Exporting high-end fish brings in foreign exchange for countries that desperately need cash. This increased income can certainly contribute to better standards of living and better nutrition, especially when combined with education, family planning, and a whole range of social development programs. Wills also believes that "all aquaculture feeds somebody somewhere, at both ends. . . . Hard currency is brought into the countries to the workers at every level—from the women in the plants to the people who drive the fish trucks along the sides of the dikes, to the feeders who feed [animals] by hand—their wages go up, their hard currency increases, and they in turn can buy food to feed their families and eat and live." On the other hand, the larger the percent of exports, the more of a net decrease in domestic consumption, and in many cases the protein source that they work so hard to harvest for other countries is absent from their own lives. Brown continues, "Nearly 1 billion people, predominantly in Asia and coastal developing countries, rely on fish for a majority of their protein needs. Over the long term, exporting a growing share of fish may result in decreasing per capita supplies of food."[60]

Looking at this practice from a global perspective, it is evident that the largest, wealthiest nations (America being one of them) are in a position to make demands on resources that can and do dramatically alter the landscape of our planet. In the United States scores of laws regulate pollution, chemical use, and employee rights. Unfortunately, those same regulations do not exist in most of the emerging nations that supply us with our food. David Corten, in his book *When Corporations Rule the World*, takes up this discussion:

> The countries that are consuming beyond their own environmental means control the rule-making process of the international economy. They adjust the rules to ensure their own ability to make up their national environmental deficits through imports, often without being mindful of the implications for the exporting countries. The pattern is most clearly revealed by looking at the export side of the equation: El Salvador and Costa Rica

grow export crops such as bananas, coffee, and sugar on more than one-fifth of their cropland. Export cattle ranches in Latin America and southern Africa have replaced rain forest and wildlife range. The lands used by Southern countries to produce food for export are unavailable to the poor of those countries to grow the staples they require to meet their own basic needs. The people who are displaced to make way for export-oriented agriculture add to urban overcrowding or move to more fragile and less productive land that quickly becomes overstressed. These dynamics are invisible to Northern consumers, who, if they do raise questions, are assured that this arrangement is providing needed jobs and income for the poor of the South, allowing them to meet their food needs more cheaply than if they grew the basic grains themselves. It makes for a plausible theory, but in practice, the only certain beneficiaries of this shift of the food economy to trade dependence have been the transnational agribusiness corporations that control global commodities trade.[61]

In addition to looking at the harm our food preferences are doing to other countries, we also have to examine how biocolonialism affects our health and well-being. Many people, myself among them, believe that buying and supporting local, regional food supplies is much more than simply supporting local economies; it is the basis for safe, clean, sustainable food. In my opinion, many of the problems with the safety of our food supplies are directly related to biocolonialism. Food produced by anonymous sources in other countries under less than acceptable conditions is no way to ensure our good health.

FOOD FOR OUR HEALTH?

Health and safety is what this is all about. Ours is a technological age that allows change to occur so rapidly that even the governmental agencies assigned to the task of ensuring our safety cannot keep up. We also live in a country that celebrates liberty and freedom of choice—but our right to free choice is impaired if we cannot make informed choices based on all available information, not just the facts and figures a company or group of companies wants us to have.

The labeling of bioengineered products is one instance in which we are being robbed of our freedom to choose. In May 1992 the FDA announced that special labeling would not be required for genetically engineered foods, saying that "unless substantial, qualitative differences exist, all foods derived from new plant varieties produced by genetic engineering would be regulated no differently from foods created by conventional means," [62] report authors Marc Lappé and Britt Bailey. Effectively, the FDA is saying that food products created with genetically altered crops are the same as those made from non–genetically engineered crops. Bioengineered foods stand on grocery store shelves, unmarked, alongside nonbioengineered foods. To date, there is absolutely no way to tell which is which, or to make judgments about the health risks that may or may not be associated with consuming these foods.

The FDA's decision touched off protests among food professionals, including the nation's leading chefs and many wholesalers and retailers. What were they so upset about? Everything from the lack of long-term studies on bioengineered products to environmental damage and allergens. In *The Biotech Century*, Jeremy Rifkin unsettlingly predicts that "human beings may yet turn out to be the ultimate guinea pigs in the radical experiment to reseed the Earth with a laboratory-conceived second Genesis."[63] If we do not know which products contain bioengineered components, it is nearly impossible for us to avoid them, protest them, or even seek them out.

Another possible drawback of the use of bioengineered products without labeling is diminished opportunity for organic agriculture. To expand the scope of organic farming, agribusiness corporations are seeking permission to use genetically engineered crops in the National Organic Program. Organic farmers stand in strong opposition. On April 17, 1998, the *Wall Street Journal* reported that Monsanto Corporation formally requested that the USDA delay its decision about including genetically engineered crops in the Organic Standards for three years. In a formal comment to the USDA, Monsanto said that the delay would allow for further discussion and education while both biotechnology and the definition of organic continued to evolve.[64] Many people involved in the National Organic Program debate believe that Monsanto wants a hiatus not for altruistic reasons, but because it believes that within three years bioengineered products will be such a part of our food supply that they will become nothing more than indistinguishable ingredients, and labeling will be a moot point.

I asked Phil Angel why Monsanto is against the labeling of bioengineered products, while the same products are being labeled in the United Kingdom. Monsanto saw the issues as two separate ones, he replied; "Monsanto's position is that labeling should be for health, safety and nutritional reasons. The mere fact of additional labeling is not necessary in a food product that has been rendered essentially the substantial equivalent of its nonengineered cousin. That is the position that is taken in this country, that is the position that has been taken in Europe, and that is the position of the federal government. However, in Europe, you have an unusual situation with regard to issues of food safety, and public confidence, primarily because of the BSE [bovine spongiform encephalopathy, or 'mad cow disease'] scandal that occurred in the UK. We have supported the position of the food retailers, who essentially said, 'We believe that our customers want to have labeled products that contain a measurable DNA or protein from a genetically modified ingredient, consumers just want to know.' "[65] In Europe, a possibly more informed public has demanded the right to know whether a product has been bioengineered so they can decide for themselves whether they want to consume it. Here in the United States we have not demanded that right to know, and if corporate agribusiness has its way, we may never have that right.

In 1996 it was proved that transgenic foods could present health hazards. Scientists at the University of Nebraska discovered that soybeans containing a Brazil nut gene caused a "potentially fatal reaction" to people with Brazil nut allergies. The soybean was never marketed, but that was not due to governmental regulations. In fact, at this time companies are not required to give the FDA information about their product. They do not even have to do studies like the one done in Nebraska. Testing is voluntary, and companies do it themselves. Marc Lappé and Britt Bailey, the authors of *Against the Grain: Biotechnology and the Corporate Takeover of the Food*, lament, "By our reckoning, the burgeoning use of transgenic food constitutes a nonconsensual experiment on a mass scale. For the first time since the Amendments to the Food and Drug Act mandated disclosure of food ingredients, people are being asked to accept and ingest foodstuffs without being told when and if their food contains engineered gene products."[66]

This lack of quantifiable research dealing with bioengineered foodstuffs may lead to problems for generations to come. Lappé and Bailey explain how: "In 1997, infant soy-based formula, using soy protein isolate powders, was used as a nutritional alternative for nearly 7 percent of all American infants. In lactose intolerant babies and infants, all or most of infants' nutritional needs are being met by exposure to transgenic products whose nutritional value and possible side effects remain untested. If any problems arise from transgenic crops, infants will likely be the first to show them."[67]

Unless we can safeguard the health and well-being of our children and ourselves, there will be no future for humanity. The following chapter describes government responsibility for the safety and longevity of our food supply, as well as what we can do to influence the choices our representatives make.

▪4▪

The Government:
The Hand That Shapes
and Molds Our Food

Since recorded time, food has been about state-building and about wealth. Rulers have justified the centralization of power through promises of protection, not only against hostile attacks, but also against famine. A major way to centralize power is to centralize control over land and labor, and the most direct link between land and labor is food. As for wealth, it is only the useful objects [that it affords] its holders. Numbers recorded in accounts and portfolios can evaporate as easily as they were entered. Bluntly, there is no wealth without food.

—Patricia Allen, *Food for the Future*

As you read in chapter 1, the U.S. government plays an important role in American agriculture. To provide safe and plentiful food supplies, laws are passed, food-related governmental departments are established, and countless subsidies are granted. The U.S. Department of Agriculture (USDA), Environmental Protection Agency (EPA), Food and Drug Administration (FDA), Department of Health and Human Services (HHS), Food and Nutrition Services (FNS), along with eight others, are the core governmental groups with responsibility for overseeing farms and farmlands, water supplies, food for children and the impoverished, restaurants, and the safety of our entire food supply.

The government influences everything we eat as well as every person who handles our food. From beef to bacon, cattlemen to fishermen, Guatemalan strawberries to Florida oranges, chemical scientists to McDonald's employees, bioengineer to restaurant dishwasher, and migrant grape picker to the CEO of Monsanto, the government presides over every facet of the food supply chain. Nothing, not even natural health food supplements, is currently exempt from some form of legislation. Surely the government's chief concern is the health and safety of the citizens of this

country, but so many other factors are involved that it is sometimes difficult to see whose interests our elected officials support. The reality is that
governmental rules and regulations often favor powerful moneyed con-
stituencies representative of the largest companies and wealthiest citizens.
In this country, you and I are encouraged to take the safety and abundance
of our food supply for granted. Perhaps we should not.

THE GUARDIANS OF OUR FOOD SUPPLIES

Everyone must eat to live, and in order to survive our bodies need healthy
nutritious food. Every day we count on other people to take care of the food
we consume. The U.S. federal government oversees twelve primary agencies responsible for the health and safety of our food supplies. According to
the National Restaurant Association, the food service industry generated
$336 billion in revenue in 1997. The numbers—of people, locations, and the
amount of product that come under the government's jurisdiction—are
staggering, especially when one considers that only $1 billion is spent annu-
ally on food safety. This is not even 1 percent of the total revenue earned.

Theoretically, government agencies are accountable for every piece of
food produced, imported, or sold in this country. In reality, this is an impossible task. Of the twelve agencies, the USDA and the FDA are primarily
concerned with food safety. Their responsibilities include the regulation and
approval of food additives, chemicals, pesticides, and other potential contaminants, testing for pesticide residues on local and imported produce, and
the supervision to ensure safe food handling, packaging, and processing. To
administer these duties, agencies appoint inspectors to positions in meat-
processing and -packing plants around the country. The safety of our beef,
poultry, and pork is in their hands.

In January 1988, Daniel Puzo addressed this issue in *Restaurants and
Institutions*: the "U.S. Department of Agriculture is responsible for inspect-
ing seven billion meat and poultry carcasses annually in 5,900 processing
plants. The USDA is vastly outgunned, especially as protein producers
increase efficiencies with high-tech equipment and state-of-the-art facilities. The Food and Drug Administration is in worse shape. According to the
Government Accounting Office (GOA), the watchdog agency of Congress,
food processing plants under FDA jurisdiction are inspected once 'every 10
years,'—a decrease in frequency from once every three to five years in
1992."[1] One of the challenges that we face, and which will be discussed in
chapter 6, is how we can require legislation that will allow for more inspectors and a more intensive inspection system.

Throughout modern agricultural history numerous chemicals, pesticides,
and fertilizers once thought acceptable for agricultural use and human
consumption have later been banned as a result of unforeseen health consequences. DDT and DES are good examples. In many cases chemicals

initially believed safe are later retested with advanced technology and found unsafe. Often the regulatory process becomes confused where departmental responsibilities overlap, and the required testing does not work or never gets done at all. Special interest concerns and lobbyists exert pressure on legislators to make laws that are beneficial to themselves but potentially harmful to the consumer. In *Earth in the Balance: Ecology and the Human Spirit*, Vice President Al Gore reports that "over the past fifty years, herbicides, pesticides, fungicides, chlorofluorocarbons (CFCs), and thousands of other compounds have come streaming out of laboratories and chemical plants faster than we can possibly keep track of them."[2] Above and beyond many of the other complications facing regulators is simply the government's inability to test every new compound and chemical as quickly as it arrives in the marketplace.

As I have been a cook all my adult life, governmental regulation of restaurants is something with which I am very familiar. The government establishes rules regarding sanitation, proper food handling, and safe temperatures for refrigeration units and hot food–holding equipment. Local sanitation officers inspect food service operations, teach classes for restaurateurs, and in some geographic areas even demand that food service employees receive sanitation certification before they can work in a restaurant. Rules are only as effective as those enforcing and following them. Inspectors must be well trained, and restaurateurs must be able to work as a cooperative part of the process in order to have a healthier food service environment.

Unfortunately, it is not uncommon for chefs and restaurateurs to cover up violations and make light of inspections. If inspectors are not particularly knowledgeable, are inconsistent in their enforcement of regulations, or can be convinced to look the other way, the consumer's health is compromised. These types of cover-ups occur in every segment of the industry, but there are so many restaurants and so few inspectors that they seem most prevalent in the restaurant industry. This issue was highlighted by a survey presented in 1998 at the twenty-second annual National Food Policy Conference: "Sixty percent of food service managers say their state has no requirements as to the number of hours required for training employees [on sanitation issues]. . . . Only three percent of food service managers believe cleanliness is an issue that must be covered in order for employees to understand safe food handling. . . . The Centers for Disease Control estimates that 79 percent of all foodborne illness results from food preparation and storage errors in food service establishments."[3]

It is not necessarily our nation's food handlers, however, who pose the greatest risks. If we continue to become bigger consumers of global food supplies, we will increase our exposure to food contamination from food handling by untrained workers. When I first became aware of this issue, I was interviewing for a job in South Africa. I asked a supervisor what his greatest challenge in the restaurant was, and he replied, "Training the native workers in sanitation." One gentleman explained that trying to train employees who sleep with their cattle and cook with their dung why hand washing is important is an impossible task. In *Spoiled: Why Our Food is Making Us Sick and What We Can Do About It*, Nichols Fox explains,

> The people who pick our green onions and strawberries and pack them into boxes and ship them to us so that we can eat them are food handlers—our food handlers. Even if we could wave a wand and make our food sterile and completely safe, as we import more and more, the problems of the developing world are going to be our problems. Those are the hands that feed you, and it might actually matter whether they are washed or not or whether they have a latrine. If we are interested in the safety of our food, then we have to be interested in the living conditions of the people who handle it. What brought that point home was this story of a trip to Guatemala, standing in the middle of a town where the number of cholera cases was growing daily, and a truck with Vermont license tags. On the side it said, "Vermont Strawberries." It was there, apparently, to pick up Guatemala's crop and to bring it home to New England so that they might have strawberries in winter.[4]

It was Guatemala that in 1997 sent us the raspberries that sickened nearly 2,500 people. By the FDA's own admission, the organism responsible, cyclospora, would not have been detected by an inspection.[5] The United States

> currently imports nearly six hundred different types of food from nearly 150 different nations. Over 6 billion pounds of fruit and 8 billion pounds of vegetables are imported into the country annually. Yet FDA tests only eight thousand imported fruit and vegetable samples yearly, which means that on average roughly one residue test is performed for each 2 million pounds of food ... one half of all winter fruits and vegetables imported to the United States come from Mexico, and this percentage may rise appreciably given the recent North American Free Trade Agreement. Mexico produces 97 percent of our imported tomatoes, 93 percent of imported cucumbers, 95 percent of imported squash, 99 percent of imported eggplant, 68 percent of imported melons, and 85 percent of imported strawberries.[6]

These facts just serve to underscore the necessity for a more localized food supply and/or stricter importation guidelines in conjunction with worker training and education.

Food-borne illness outbreaks are on the rise in the United States, and our government's position is often reactive, not proactive. The media is largely responsible for the government's post-outbreak heightened awareness. Hard-hitting media stories often spur local health departments to action. In 1998 the *Orlando Sentinel* published an article exposing the lack of disciplinary action taken against restaurants with poor or failing inspection scores. Not long afterward, the Florida Restaurant Association requested additional state health inspectors. In Los Angeles, a KCBS-TV news series prompted a handful of significant reforms, including a countywide ordinance requiring restaurants to "hang A, B, or C letter grades on their front doors to reflect inspection scores." San Francisco followed suit, drafting legislation "calling for the food-safety training of all restaurant shift managers by Jan. 1, 2000." The President's Council on Food Safety, introduced in 1998, promises to help keep consumers better informed and is working on developing a strategic plan for federal food safety education and training programs. The council's efforts will be the result of collaboration among the secretaries of Agriculture, Commerce, Health and Human Services, and the Environmental Protection Agency, as well as the assistants to the president for science and technology and domestic policy.[7]

No matter how diligently the government works to enforce regulations, and no matter how vigilant chefs and restaurateurs are, it is ultimately the customer who has to take responsibility for food safety. When consumers vote with their buying dollars, the industry will respond. While it is a sad commentary on our society, the threat of litigation is an effective method to promote improved food-handling practices.

A Jack-in-the-Box restaurant located in the state of Washington was the scene of one of the most widely publicized *E. coli* outbreaks in recent memory. Multiple lawsuits were filed, and Jack-in-the-Box was forced to pay. Exact settlement amounts were not made public, but it is clear that both Jack-in-the-Box and the packing company that distributed the tainted meat were held liable. Beef suppliers did agree to pay the parent company of Jack-in-the-Box $58.5 million in a settlement.[8]

Unfortunately, the list goes on. Following the Jack-in-the-Box catastrophe, similar cases of food contamination have led to sickness, death, and litigation. Odwalla juice is another example. Odwalla agreed to pay a $1.5 million fine for selling the apple juice that killed a sixteen-month-old girl and sickened seventy other people in 1996. In Florida, Bauer Meat Co. was shut down by the USDA, which stated that *E. coli* contamination made the product "unfit for human consumption." The company is also a target of investigation by the USDA's Office of the Inspector General, and has been sued by one of the children sickened by eating meat from his school's cafeteria.[9] Although *E. coli* is on the rise—reported cases were up over 100 percent between 1994 and 1998, it is too early to measure the impact of litigation on food safety standards. The question is, How many illnesses or how many deaths or how many lawsuits will it take for our elected officials to make food safety a national priority?

Becoming an informed consumer has never been easier. Reports stating a restaurant's cleanliness and food safety ratings, though available to the public, have traditionally been the property of the government and restaurant. With the accessibility of government documents on the Internet, however, consumers may now read exactly what inspectors say about facilities and the kinds of infractions that they have incurred. What is good for the consumer in this case may be what motivates the restaurateur to becoming more vigilant about safe food-handling practices.

We do have a system of checks and balances that ensure the right to justice goes both ways. For example, the law provides a vehicle for the food industry to sue anyone who publishes false information raising safety concerns about agricultural products. One of daytime TV's talk show divas, Oprah Winfrey, was hauled into court in 1998 for allegedly libeling the cattle industry. These so-called food disparagement laws may actually make it more difficult for the media to bring to light their concerns about inadequacies in the industry. Winfrey won her case in a much-publicized battle that ended with a court ruling that Winfrey had the right to publicly express her fear of hamburger. It is unclear what effect Oprah's win will have on the industry in the long term, but it once again brought debate to the public forum.

The government, the restaurant industry, food producers and processors, and the transportation industry agree that food safety is of immediate and growing concern. All have worked tirelessly to make it a top priority. The American Culinary Federation has made food safety and sanitation training a mandatory part of its certification process, and the Culinary Institute of America as well as the majority of professional culinary schools integrate the same type of training as a required part of the degree process. The National Restaurant Association has instituted its own intensive industry-training program. Known as Servesafe, it has graduated over 800,000 food service managers since its inception. In a similar move, the Dietary Managers Association created the Certified Food Protection Professional (CFPP) certification, available to all food service professionals.[10] It is clear that the majority of food professionals understand their responsibilities in keeping our food safe and clean. Not only has the food service industry taken sanitation issues seriously, but in the past couple of decades meat, poultry, and seafood producers and processors have all taken steps to ensure a safer food supply. Government-mandated Hazardous Analysis Critical Control Point (HACCP) certification for the meat and seafood industries, as well as education and training throughout the industry, has had a positive impact on many food producers and handlers nationwide. Many food service industry producers, such as Gail Allen, president of SYSCO Foodservice Albany, and Steve Connolly, president of Steve Connolly Seafood, tell me that HACCP certification has had a positive impact on all segments of their businesses. They both believe the regulations have made them more aware of potential hazards not only in their own companies, but also in their transportation systems and in their suppliers' operations. I believe that most successful nationwide suppliers have also used HACCP to their benefit. For me, as a purchaser, discussing a company's HACCP plan is a prerequisite to doing business.

FDA: ADDITIVES, CHEMICALS, AND FOOD LABELING

The Federal Food, Drug and Cosmetic Act of 1938 (FFDCA) gave the FDA the authority to regulate food additives. Examples of additives include food coloring, thickeners, sweeteners, or any other item that is not naturally part of the product. Through premarket testing, the FDA determines the level of risk to human health as part of the approval process. The agency specifically regulates the type of food in which an additive can be used, the maximum quantity that can be used, and the required additive labeling information. Laws governing food additives include the Delaney Clause, which bans the addition of any carcinogenic products. The clause reads, "No additive shall be deemed to be safe if it is found to induce cancer when ingested by man or animal, or if it is found, after tests which are appropriate for the evaluation of the safety of food additives, to induce cancer in man or animal."[11]

There are, however, problems with this law. First, products are rarely tested for their long-term effects on the human body. We only discover what harm may be done through continued consumption of a certain food additive or preservative. There are always those who speculate about the potential shortcomings of a new product, such as those who believe that in the not-too-distant future Olestra will cause 50,000 new cancer cases a year, but until this is definitively prove, an additive like Olestra will remain on the market. Second, testing is primarily done on adults; children, who may be at the greatest risk, are most often not considered. In actuality, it has been shown that 40 percent of cancer is diet related. I stand by the point I made earlier: our government is more often than not reactive rather than proactive. Only when enough people die from diseases that are related to additives does the government take action on behalf of consumers.

Food labels now only incompletely disclose the chemicals present in a product. For instance, the FDA does not require manufacturers to list pesticide or other chemical residues remaining on or in products postharvest or after final processing. In many cases, consumers have absolutely no way of knowing which chemicals were used to produce their food. Alar, which gained widespread use in the late 1980s, is a good example. Most Americans had no idea that Alar was being used to increase apple production in many orchards. In 1990, after it was found to be potentially hazardous—especially among children, who are the primary consumers of apple products, such as apple juice and cider—it was removed from the market-

Governmental Food Regulatory Agencies

The government employs so many agencies to police our food supplies that it is impossible to remember who they are, what their jurisdictions are, and how they relate to each other. Explanations of each and every one could be a book in its own right, but the following synopses provide the basic understanding needed by every consumer. Complete contact information is included in appendix A, p. 227.

The *Department of Health and Human Services* (HHS) is the U.S. government's principal agency for protecting the health of all Americans. Among the agency's more than 300 programs are those that work toward the prevention of infectious disease and those that assure food and drug safety. The Food and Drug Administration (FDA), the National Institutes of Health (NIH), and the Centers for Disease Control and Prevention (CDC) fall under the umbrella of the Department of Health and Human Services.

The mission of the *National Institutes of Health* (NIH) is to uncover new knowledge that will lead to better health by conducting research in its own laboratories; supporting the research of nonfederal scientists in universities, medical schools, hospitals, and research institutions throughout the country and abroad; helping in the training of research investigators; and fostering communication of biomedical information to the public. The NIH is one of eight health agencies of the Public Health Service, which, in turn, is part of HHS. Comprised of twenty-four separate departments, NIH had a budget of more than $15.6 billion in 1999.[1]

The *Centers for Disease Control and Prevention* (CDC) is responsible for promoting health and quality of life by preventing and controlling disease, injury, and disability. The CDC has eleven centers whose goals are the prevention and spread of communicable diseases as well as to monitor and prevent food-borne illness outbreaks and their dissemination within the public. The CDC staff includes approximately 6,900 employees in 170 locations.

The *Food and Drug Administration* (FDA) touches the lives of virtually every American every day, says the FDA website. It is the FDA's job to see that the food we eat is safe and wholesome, that it is labeled truthfully, and that the feed and drugs given to farm animals are safe both for the animals and for the people who consume animal products. The FDA strives to ensure our safe food supply with over 1,000 inspectors who visit more than 15,000 facilities a year and supervise both nationally produced food products and imports from a total of almost 95,000 businesses.[2]

The U.S. Department of Agriculture (USDA) is one of the most all-encompassing agencies that regulate our food supplies. It has numerous departments and tens of thousands of employees in offices throughout the country. The mission of the USDA is to enhance the quality of life for the American people by supporting the production of agriculture in the following ways:

* ensuring a safe, affordable, nutritious, and accessible food supply;
* caring for agricultural, forest, and range lands;
* providing economic opportunities for farm and rural residents;
* expanding global markets for agricultural and forest products and services;
* working to reduce hunger in America and throughout the world.

Governmental Food Regulatory Agencies (continued)

The following are the most prominent of the agencies under the auspices of the USDA that govern our nation's food supplies:

The *Farm and Foreign Agricultural Services* is responsible for administering agricultural price and income support programs, production adjustment programs, and the Conservation Reserve Program. The Federal Crop Insurance Corporation programs are administered there, as are the farm lending programs for agricultural producers and others engaged in production of agricultural commodities.

The *Farm Service Agency* (FSA) helps stabilize farm income, aids farmers in conserving land and water resources, provides credit to new or disadvantaged farmers and ranchers, and helps farm operations recover from the effects of disaster. FSA runs commodity loan programs for wheat, rice, corn, grain sorghum, barley, oats, oilseeds, tobacco, peanuts, upland and extra-long-staple cotton, and sugar.

The *Food and Nutrition Services* (FNS) administers the fifteen nutrition assistance programs of USDA. These programs, which serve one in six Americans, represent our nation's commitment that no one in our country should fear hunger or experience want. FNS administers the following food assistance programs:

The *Food, Nutrition and Consumer Services* (FNCS) ensures access to nutritious, healthful diets for all Americans through regulation, education, and inspections.

The *Food Stamp Program,* the cornerstone of the USDA nutrition assistance programs, served an average of 22.9 million people each month in 1997.

The Special Supplemental Nutrition Program for Women, Infants and Children (WIC) improves the health of low-income pregnant, breast-feeding, and non-breast-feeding postpartum women, and infants and children up to five years old.

The WIC *Farmers' Market Nutrition Program* provides WIC participants with increased access to fresh produce. WIC participants are given coupons to purchase fresh fruit and vegetables at authorized local farmers' markets.

The *Commodity Supplemental Food Program* is a direct food distribution program with a target population similar to WIC's, and it also serves the elderly.

The *National School Lunch Program*, the *School Breakfast Program*, the *Special Milk Program,* and the *Child and Adult Care Food Program* are all under the auspices of FNS and are designed to assure the nutritional health of school-age children and disabled adults.

The *Nutrition Education and Training Program* (NET) supports nutrition education in the food assistance programs for children through Team Nutrition and other school programs.

The Center for Nutrition Policy and Promotion coordinates nutrition policy in USDA and provides overall leadership in nutrition education for consumers. The center is the link between basic science and the consumer and coordinates and disseminates the Dietary Guidelines for Americans.

The *Food Safety and Inspection Service* (FSIS) conducts specific activities to ensure the safety of meat and poultry products consumed in this country, assuring that the nation's meat and poultry supply is safe, wholesome, unadulterated, and properly labeled. USDA inspectors and veterinarians conduct slaughter inspection of all carcasses at meat and poultry slaughtering plants for disease and other abnormalities and sample for the presence of chemical residues.

Governmental Food Regulatory Agencies (continued)

The *Agricultural Marketing Service* (AMS) programs enhance marketing and distribution of agricultural products and assure a competitive, fair, and cost-effective marketplace by disseminating market news, developing grade standards, and providing commodity inspection, grading, classing, and certification. They are also responsible for developing organic standards, improving direct marketing by farmers to consumers, and providing technical assistance on rural transportation issues.

The *National Resources and Conservation Service* (NRCS) is the federal agency that works hand in hand with the American people to conserve, improve, and sustain natural resources on private lands. The standards for conservation systems address such areas as erosion control, animal waste management, irrigation water management, wetlands conservation and restoration, flood control, streambank stabilization, and a plant materials program that introduces new ways to use plants for revegetation, land stabilization, and landscape enrichment.

The Research, Education, and Economics' mission is to develop cutting-edge technologies that improve food and fiber production and enhance the safety of the national food supply. USDA research finds many new uses for the nation's agricultural bounty, improves crop varieties, and prevents crop losses and animal diseases caused by various pests and pathogens.

The *Agricultural Research Service* (ARS) provides access to agricultural information and develops new knowledge and technology needed to solve technical agricultural problems of broad scope and high national priority. The goal is to ensure an adequate supply of high-quality, safe food and other agricultural products to meet the nutritional needs of consumers, to sustain a competitive food and agricultural economy, to enhance quality of life and economic opportunity for rural citizens and society as a whole, and to maintain a quality environment and natural resource base.

The mission of the *Environmental Protection Agency* (EPA) is to protect human health and to safeguard the natural environment—air, water, and land—upon which life depends. The EPA's purpose is to ensure that all Americans are protected from significant risks to human health and the environment where they live, learn, and work.

The National Marine Fisheries Service (NMFS) or "NOAA Fisheries" is a part of the National Oceanic and Atmospheric Administration (NOAA). NMFS administers NOAA's programs, which support the domestic and international conservation and management of living marine resources. NMFS provides services and products to support domestic and international fisheries management operations, fisheries development, trade and industry assistance activities, enforcement, protected species and habitat conservation operations, and the scientific and technical aspects of NOAA's marine fisheries program.

Source: U.S. Government.

1. National Institutes of Health, website, www.nih.gov/icd/programs.html.
2. www.fda.gov/opa.com.

place. Millions of children consumed Alar year after year before it was banned from use. In the decade that has passed since Alar came into the public eye, it has never been proved to cause specific ailments, but many have come to see media attention to Alar as the consumer's first wake-up call on the issue of food and chemical safety.

As consumers of food products, we are largely ignorant about the pesticides used to grow our food because we trust that government agencies like the FDA are doing a thorough job of monitoring additives for us. The reality is that in our global marketplace it is the responsibility of the consumer to be educated about these issues. Did you know, for instance, that when you go to the supermarket to buy "fresh" scallops, you may actually be getting shellfish containing up to 25 percent water in a solution with sodium tripoly phosphate? Tripoly phosphate is a food-grade preservative that adds salt (sodium) to the fish and changes the flavor and texture of the product when cooked. Besides perpetuating a misrepresentation of the truth, the FDA could indirectly be contributing to someone's health problems by not requiring that this kind of information appear on end-user product labels. "Fresh" chickens may also contain up to 8 percent water and can be labeled as fresh even if they have been chilled to 26°F—well below freezing. Additionally, under FDA guidelines a product labeled "fresh" may be doused with pesticides just before or just after harvest; produce may be cleaned with a chlorine or acid wash; and raw products may be treated with ionizing radiation. All that, and the label would simply read "fresh."

Bioengineered products are also the FDA's responsibility, and while food additives are subjected to the "zero-based" tolerance rule outlined in the Delaney Clause, bioengineered products are handled very differently. In 1992 the FDA stated that it would focus on the safety of the end product and

Federal Agencies Responsible for Regulating, Monitoring, or Grading Various Food Industries—USDA

Food Industry	FDA	AMS	FGIS*	FSIS	EPA	NMFS
Dairy	*	*			*	
Eggs/egg products	*			*	*	
Fruits/vegetables	*	*			*	
Grains/rice	*		*		*	
Interstate conveyances	*					
Meat and poultry		*		*	*	
Restaurants	*					
Seafood	*				*	*

Source: U.S. General Accounting Office.
* Federal Grain Inspection Service.

not the process by which it was developed. Under this policy, the foods created using biotechnology are to be treated no differently than ordinary food. If, for instance, a vegetable has been bioengineered to include a fish gene that will protect it from adversely cold temperatures, the FDA will consider the food safe based on the safety of the vegetable's nonbioengineered cousin. The "new" vegetable/fish gene combination is irrelevant to the testing process.

Testing of new products is voluntary. There is no governmental requirement that all bioengineered products must be tested before being introduced to the marketplace. New products are considered as safe as they were in their original form, and may be presented to the public without labeling. Because of this, potential criticism of the "new" product is automatically disallowed. On top of that, decisions have recently been made that exempt bioengineered food from labeling. Advocates of an organic and sustainable food supply are distressed by this turn of events; ultimately, it means that bioengineered foods can become part of our food supply without our knowledge or consent. Because the long-term effects of eating bioengineered products are still unclear, we are essentially becoming human test cases without the freedom to choose. The reasons? Money and power.

There is a revolving door between government officials and the biotech industry. The FDA's no-labeling policy was written by Michael Taylor, a lawyer representing Monsanto. In 1991 he became a deputy commissioner of policy for the FDA, and after they approved rBGH Taylor moved to the USDA as director of food safety. He eventually went back to work for Monsanto as head of public policy. Food and agribusiness corporations contributed more than $71 million to federal congressional campaigns between 1989 and 1998 (reported by Common Cause). Monsanto, DuPont, Dow Agro-Sciences, and thirty-two other pesticide manufactures formed a lobbying group, Responsible Industry for a Sound Environment (RISE), which spent more than $15 million in 1996 to employ 219 Washington lobbyists.[12] This incestuous sharing of employees in combination with the power of enormous profits clearly accounts for part of the labeling issue.

One of most widely recognized—though not yet commercially adopted—bioengineered foods to recently appear on the scene is the Flavr Savr tomato. The Flavr Savr tomato has been grown and marketed in some parts of the world, but never really took hold in this country, not because of any outcry against it but because it did not taste good and lacked market appeal. Monsanto has since decided to use the technology on more flavorful types of tomatoes and see if they are more readily accepted. Fewer people know, however, about some of the bioengineered products that have been approved and are currently part of our food supply. Bioengineered soy, rapeseed (canola oil), and corn are already part of our food supply, and they are not labeled in any way. The aim of Monsanto and other companies that market bioengineered seeds is that these products will become such a ubiquitous part of the food supply that future labeling laws will have little or no impact.

When you stop and consider that we all need food to survive, what choice will consumers have if the majority of affordable food products are produced through bioengineering? Why else would Monsanto request during the Organic Standards debate a moratorium on decision making as it applied to labeling bioengineered products? Consumers must become educated and involved in these decisions and make their opinions known to agribusiness companies and government agencies.

One of the first major bioengineered products to find a widespread market niche was bioengineered rennin, approved in 1990. Rennin is a milk-clotting substance naturally found in the stomach lining of ruminants (animals like cows, sheep, and goats, which have multiple stomachs and chew cud), and is used to produce cheese. The bioengineered variety is currently being used to produce 65 percent of all U.S. cheeses. Many vegetarians celebrate bioengineered rennin as a way of making cheese without using additional animal products; however, there are others who oppose the bioengineered enzyme, believing that there has not been adequate testing of its long-term effects on humans. As I already mentioned, we consume a high percentage of bioengineered soy and corn products. We may be ingesting rBGH without knowing it, and many of us are voluntarily consuming huge quantities of Olestra. It is possible that none of these things will pose health problems for consumers, but right now we truly do not know or understand their long-term health effects.

The organic/sustainable community believes that nonbioengineered foods are safest, but is primarily concerned with giving consumers the option to make informed choices. The biotechnology sector does not want to let this happen. In fact, the industry in the United States is so adamantly against labeling that it is safe to assume the real fear is that consumer knowledge would lead to a serious attack on the bottom line. Fighting to keep unidentified bioengineered products in the marketplace is investment protection for companies like Monsanto. The only way the average consumer can fight big business is by working to stay informed.

PESTICIDES AND CHEMICALS

The question is whether any civilization can wage relentless war on life without destroying itself, and without losing the right to be called civilized. Insecticides are not selective poisons; they do not single out the one species of which we desire to be rid. Each of them is used for the simple reason that it is a deadly poison. It therefore poisons all life with which it comes in contact: the cat beloved of some family, the farmer's cattle, the rabbit in the field, and the horned lark out of the sky.

—Rachel Carson, *Silent Spring*

Rachel Carson issued her warning in 1962. In 1995 U.S. farmers used 565 million pounds of pesticides. Many say that chemical pesticides are essential to maintain productivity and keep food prices low—but a study by David Pimentel, an entomologist at Cornell University, concluded food prices would increase less than 1 percent if U.S. farmers cut their use of chemical pesticides in half.[13] One reason prices would hold steady, experts believe, is

DDT: From Miracle Cure to Deadly Killer

DDT was introduced to a world plagued by insect-borne disease and torn by global warfare. As an insecticide, its ability to "sanitize" huge areas of the landscape seemed miraculous. First isolated in 1874, DDT was recognized as a nerve poison for insects in 1939 by the Swiss Nobel Prize–winning chemist Paul Müller, and was initially utilized heavily during World War II for pre-invasion spraying. Malaria was so prevalent in the jungles that pre-invasion spraying was necessary to keep soldiers healthy enough to fight the war. At the same time, DDT was being used throughout the world to combat yellow fever, typhus, elephantiasis, and other insect-vectored diseases. The end of World War II did not bring an end to the use of DDT; on the contrary, it enjoyed global popularity as an insecticide as stockpiles of the chemical were quickly redirected toward the protection of the general population from insect-borne disease as well as toward the agriculture industry as crop protection. DDT was not the only chemical of its kind to come from the war effort. Others included benzene hexachloride (BHC), the herbicide 2,4-D, and dieldrin. Rachel Carson reported that dieldrin was five times more toxic than DDT when swallowed, but forty times as toxic when absorbed through the skin. Interestingly, dieldrin gained popularity when insects began to develop a resistance to DDT. People who had exposure to dieldrin experienced a variety of side effects, including severe seizures that continued to occur as long as four months after ingestion.

Following the war, the United Nations set up an initiative to distribute the new wonder chemicals to countries in need through the World Health Organization (WHO). As documented by author John Wargo, "By 1970, malaria had been eliminated from Europe; the Asian portion of the Soviet Union; some Middle Eastern states; much of North America (including all of the United States); major portions of Mexico and most of the Caribbean nations; the far northern and southern portions of South America; Japan; Singapore; Korea; Taiwan; and Australia. By 1970, worldwide, an area inhabited by nearly 700 million people that had been plagued by malaria in 1950, was virtually free of the disease. Many lives were likely saved by the effort. In addition, agricultural productivity rose and general public health improved."

Rachel Carson's crusade against pesticides and other agricultural chemicals began in the early 1960s. For the first time, the public was introduced to the hazards these chemicals presented to humans and animals. A new war began, and both sides launched aggressive media campaigns. One ad showed a mother spraying that hundreds of pest species have already developed resistance to one or more pesticides. As a result, chemical applications are largely insufficient protection against insects, and, as Jenny Tesar expalins in *Food and Water: Threats and Shortages and Solutions,* pesticide usage may actually serve to increase the dependence on chemicals: "Indeed, studies suggest that farmers who depend heavily on pesticides are fighting a losing battle. In 1945, U.S.

DDT (continued)

powdered insecticides right over her baby's crib. The benevolent mother smiled as chemicals showered down upon her child's body. In *Silent Spring* Carson staged her own battle:

> Perhaps the myth of the harmlessness of DDT rests on the fact that one of its first uses was the wartime dusting of many thousands of soldiers, refugees, and prisoners, to combat lice. It is widely believed that since so many people came into extremely intimate contact with DDT and suffered no immediate ill effects the chemical must certainly be innocent of harm. This understandable misconception arises from the fact that—unlike other chlorinated hydrocarbons—DDT in powder form is not readily absorbed through the skin. Dissolved in oil, as it usually is, DDT is unquestionably toxic. If swallowed, it is absorbed slowly through the lungs. Once it has entered the body it is stored largely in organs rich in fatty substances (because DDT itself is fat-soluble) such as the adrenals, testes, or thyroid. Relatively large amounts are deposited in the liver and kidneys.[1]

Mammary glands, also largely composed of fat, act as storage spaces for DDT and DDE, both of which are believed to be potential causes of some breast cancer in women.

The main concerns with DDT are that it can be passed through the food chain, may have a shelf life of decades beyond its initial usage, and can actually become more concentrated during food processing. Carson explains how this occurs:

> For example, fields of alfalfa are dusted with DDT; a meal is later prepared from the alfalfa and fed to hens; the hens lay eggs, which contain DDT. Or the hay, containing residues of 7 to 8 parts per million, may be fed to cows. The DDT will turn up in the milk in the amount of about 3 parts per million, but in butter made from this milk the concentration may run to 65 parts per million. Through such a process of transfer, what started out as a very small amount of DDT may end as a heavy concentration. Farmers nowadays find it difficult to obtain uncontaminated fodder for their milk cows, though the Food and Drug Administration forbids the presence of insecticide residues in milk shipped in interstate commerce.[2]

It was issues such as these that forced DDT from the U.S. market. Although DDT was banned from use in this country, it is still exported to other nations, where it is used to grow food products for U.S. consumption. We continue to eat foods contaminated with DDT to this day.

1. Rachel Carson, *Silent Spring* (New York: Houghton Mifflin, 1962), 20–21.
2. Ibid., 22–23.

farmers lost 3.5 percent of their corn crop to insects. By 1988, pesticide use was 1,000 times greater and farmers lost 12 percent of their corn crop."[14] When we learn to understand and appreciate that nature always finds a way to perpetuate life simply by mutating bugs to adapt to pesticides, we are in a position to make informed choices about the food we select for ourselves. There are no simple solutions here. Agribusiness profits from pesticides and biotechnology. Chemical companies like Monsanto have tremendous lobbying power. Unless these companies achieve the same level of profit through other means, they are not going to forgo products that they develop and sell in the marketplace. Like any other industry, whether telecommunications or electronics, when core business activities are under threat, business focuses its money and influence on Capitol Hill.

The use of agricultural chemicals, in the United States as well as the other countries that supply us with food, appears to be out of control. In *Our Children's Toxic Legacy: How Science and Law Fail to Protect Us from Pesticides,* John Wargo tells us that

> between 1964 and [the end of the 1990s], pesticide use in the United States doubled from 500 million to over 1 billion pounds per year. In 1983 there were approximately fifty thousand separate compounds registered by USDA; this number was reduced to approximately twenty-one thousand compounds, of which nearly nineteen thousand of these products need to be registered by EPA. The quality of data supporting the original claims of environmental health and safety were suspect and hence the need for new testing.[15]

Our technology and biology industries were discovering and identifying cures for our industrial ills much faster than the government was able to identify potential product hazards and inform the public. Wargo goes on to say,

> The EPA's inability to handle all these data was demonstrated in 1991 when a train derailed in Northern California, spilling the herbicide metam sodium into the Sacramento River, which feeds Lake Shasta. Although the manufacturer had submitted data to EPA four years earlier demonstrating that metam sodium caused birth defects, EPA did not know that it had this study, had not reviewed it, and therefore was incapable of providing appropriate warnings to pregnant women who might drink from the river, which supplies public drinking water.[16]

In 1996 many consumer advocates saw the Food Quality Protection Act as the beginning of a new era in EPA pesticide management. The law placed the burden of proof of safety on the product manufacturer, who was required to show with "reasonable certainty" that aggregated exposure to the pesticide would pose "no harm."[17] Consumer advocates had high hopes for the new law because previously the burden of proof had been in the government's hands to positively prove harm, which was almost impossible to do without tremendous amounts of testing. The new law transferred this burden of proof to the manufacturer, who now had to demonstrate, to the government through extensive testing that absolutely no harm could occur. Unfortunately, it has been the source of more problems than solutions. The Delaney Clause of the pesticide law stated a "zero cancer risk policy," but Delaney had never been enforced because of the government's inability to prove positive harm.

The Food Quality Protection Act mandated that "zero tolerance" had to be enforced, and, if strictly enforced, would ban numerous pesticides of high economical value to the agricultural industry. Farm and pesticide lobbyists went to Congress to ward off this potential blow to the industry. But problems with the language in the Quality Protection Act eventually came to mean that states would be prevented "from adopting more stringent pesticide regulations than those set by EPA."[18] In other words, a community would not be allowed to ban a chemical allowable under EPA regulations. Once again, communities that might wish to enforce a higher level of protection than the government mandates are not permitted to do so.

The Environmental Protection Agency (EPA) is responsible, as its name suggests, for protecting the environment. Created by the Nixon administration in 1970, the EPA was a consolidation of personnel from thirteen different federal agencies, including divisions from within the Departments of the Interior, Agriculture, Health, Education, and Welfare. Since its establishment in 1970 the EPA has set water quality standards, regulated regional water pollution, and been responsible for protecting the public from radiation. The EPA also regulates the handling and control of chemicals deemed hazardous to public health. It is in this capacity that it oversees the use of pesticides and chemicals in agricultural food production. The agency works hard to be effective in protecting us and our environment from harm, but like all other large government bodies it suffers from politics, and lack of funding.

The foundation for most of the EPA's pesticide policies was generated through intensive debate among the agency's inherited staff members as well as the pesticide lawsuits passed along from the USDA. One of the EPA's basic laws, the Federal Insecticide, Fungicide, and Rodenticide Act, actually directs the EPA to take the potential economic impact of a product into consideration during the approval process. Where only consumer health and safety should have been concerns, lobby efforts and politics come into play because if undue financial burden can be proved to cause harm to industry, a chemical may not be removed from the market. Additionally the agency relies on data and analysis compiled, not by its own scientists but by those of the very companies it regulates.

The EPA determines risk factors by weighing the benefits of chemicals against their potential risks. For example, a chemical may be deemed hazardous to public health, even classified as a possible carcinogen with potential to attack the nervous system or cause birth defects and sterility, and still be within acceptable EPA guidelines. In fact, according to John Wargo, the "EPA and USDA before it, set nearly ten thousand separate tolerances for different pesticide residues in different foods. Some foods, such as apples and milk, are allowed to contain residues of nearly one hundred separate pesticides."[19] This may seem inconceivable, but if the benefits of chemical use include higher yield or less potential loss, the EPA might decide the benefits outweigh the risks. Even though the EPA determines risk factors, the FDA regulates chemical presence in food. The two government agencies work separately with different sets of rules, both endeavoring to assess the potential risks and benefits associated with chemicals and our food supply. This inevitably leads to compromises that the consumer may find less than agreeable. For instance, Michael Ableman writes,

> although pesticide residues were found on forty-eight percent of all food sampled by the FDA between 1982 and 1985, the food supply was still considered to be "safe." The process seems even more frighteningly lacking when we consider that the FDA examines only about one percent of the food supply. Its most sophisticated testing methods are currently able to detect the presence of fewer than half of the almost 654 pesticides, herbicides, and fungicides currently in use. What's more, the EPA has never studied the health effects when several chemicals have been combined on a single crop—a commonplace occurrence.[20]

The EPA is also responsible for managing pesticide contamination in our ground and drinking water. Forty percent of the U.S. population depends on underground water for consumption; 90 percent of all rural residents are dependent upon this resource. In spite of such heavy reliance on groundwater, it was not until 1979 that the government conducted a study highlighting the hazardous compounds found in our water supplies. The EPA found sixteen pesticides in groundwater by 1985, and by 1988 forty-six were evident—all from normal agricultural use.[21] In 1998 the EPA unveiled new regulations that would, for the first time, provide consumers with details on the chemicals found in their drinking water and whether their water meets federal standards. Water agencies, which oversee the water districts that supply water directly to the consumer, are now required to inform consumers of water sources, contaminants found in the water, and whether those contaminants are present in levels that exceed EPA health standards. In addition, water agencies must report potential health risks associated with drinking contaminated water, and whether there have been any violations by or enforcement actions against their water agencies in the preceding year.

Because tens of thousands of potentially hazardous agricultural compounds are licensed for use without comprehensive testing, the consumer is at a tremendous disadvantage when it comes to information. While it is easy to empathize with the volume of work facing the mere mortals who staff the EPA, it is also obvious how the agricultural industry exerts enormous pressure on lawmakers. Financial gain is the responsibility companies have to their shareholders. Consumers who own stock in publicly traded chemical and agribusiness companies should not be surprised by intensive lobbying efforts. I believe that it is a disturbing commentary on our society that people know more about the financial health of a company than they know about their personal health and the relationship between diet and disease.

BEYOND PESTICIDES: IRRADIATION AND OUR FOOD SUPPLIES

Irradiation is a relatively new method of food preservation; this controversial process was discovered in the 1940s and tested for years by the army. It delays ripening of fruits and vegetables; inhibits sprouting in bulbs and tubers; disinfects grain, cereal products, fresh and dried fruits, and vegetables; and destroys bacteria in fresh meats. Long used as a way to prevent shrinkage due to insect infestations, irradiation has become commonplace for dry herbs and spices. Although testing suggests that irradiation may help reduce agricultural reliance on pesticides, it has, to date, been strongly opposed in the marketplace.

In the 1990s, with the explosion of media attention on food-borne illnesses like *E. coli* (a pathogen that can survive both freezing and heating to temperatures up to 150° F), irradiation once again came to the fore as a method of food sterilization. Proponents believe irradiation kills microorganisms and extends shelf life, allowing for lengthier storage of fresh meat, produce, and fruit. Spices are currently the only food products being regularly irradiated on a commercial basis. Consumers may be surprised, however, to learn that a number of other items are exposed to radiation before they hit supermarket shelves, including cosmetics, and tampons and other medical supplies. With FDA approval we may soon see poultry, pork, vegetables, and fruits added to that list.

A National Restaurant Association study done in 1996 revealed that one of the dining public's top ten concerns is food safety, and lately it seems that food-borne illness outbreaks or product recalls occur on a weekly basis. With public fear at an all-time high, the stage has been set for the food service industry to advocate the use of irradiation as a method of ensuring food safety. In fact, the FDA has recently approved irradiation as a safe and effective method of food safety in meat processing, though as of this writing it has yet to be used commercially. According to a 1997 article in *Nation's Restaurant News*, "Consumers spooked by visions of glowing steaks should have no fears; after three years of review, the agency [FDA] proclaimed that irradiation was a safe and effective way to rid meat of disease-causing microbes without compromising nutritional qualities."[22] Consumers and consumer advocates remain skeptical of the technology, and many chefs oppose irradiation as well. Chefs are concerned about changes in flavor and texture, and consumer advocates continue to worry that a product sterilized by irradiation may be more vulnerable to contamination.

I spoke to Fred Genth, at the time executive chef and director of product development for Monfort Beef, about irradiating beef as a method of pathogen control. Monfort, one of the three largest beef producers in the United States, slaughters approximately 25 percent of U.S. beef, and is extremely concerned with methods of eradicating food-borne diseases. Fred told me, speaking on his own behalf and not as a representative of the company, that he is not in favor of irradiation; in his opinion it changes the flavor of the beef. He went on to tell me that the larger issue involves the potential for cross-contamination after the product is irradiated. He explained, "[We] are left with a very sterile product, which can be re-contaminated very easily, down the line, after it leaves our hands. And that is the area that we are not sure about, or one of the areas we are not sure about. [We have no idea] how bad that recontamination could be. I think there are better ways."[23] Research and testing regarding recontamination has yet to be done, but at the same time, the USDA has echoed the FDA's approval of irradiation on foodstuffs.

The Chicken and the Egg

As with *E. coli,* salmonella reached a heightened level of public awareness in the 1990s. In fact, it has become much more prevalent in the final two decades of the century as a result of an explosion in consumer demand for numerous products, especially chicken and eggs.

Like any business, the poultry industry responded to increased consumer demand by enlarging and intensifying production. The consequence, however, is that more animals are raised in smaller spaces, where disease is more easily passed from animal to animal. Ease and speed of transport have also contributed to the contamination problem by distributing infected animals throughout the country, making it more difficult to identify infected flocks and stop the spread of disease. Primarily associated with chicken meat, salmonella has more recently been found in chicken eggs. In *Spoiled: Why Our Food Is Making Us Sick and What We Can Do About It,* Nichols Fox states, "Because salmonella managed at some point to get into the ovary of the chicken to produce eggs prepackaged with pathogens, and because changes in how we produce eggs have actually encouraged this microbe, of the hundreds of uses of the egg, only those in which it is thoroughly cooked are now considered completely safe."[24]

Between 1985 and 1991, 380 outbreaks of salmonella were reported by thirty-seven state health departments. This prompted the FDA to "redesignate the egg as a hazardous food under its model food codes" in 1990; "Eggs would have to be refrigerated, as were other foods of animal origin, even during transportation."[25] The industry fought the FDA, especially resisting refrigerated transportation, and many USDA rules were flagrantly disregarded by the egg industry. Public outcry prompted the USDA to take action in the aftermath of the late 1980s' "epidemic of salmonella"; nevertheless, when the agency became seriously concerned and attempted to enforce its rules upon the industry, this change in status and regulation had little effect on the number of outbreaks. In fact, in 1996 the FDA reported that the new rules had never been followed because the egg industry believed they were too restrictive. In an attempt to get around the problem of salmonella in egg products, chicken farmers whose flocks are known to be infected with the disease will voluntarily reroute their eggs to the "breakers," or processing plants that mechanically crack and pasteurize liquid egg product. The pasteurized eggs are then sent to institutions, other food processors, and restaurant kitchens.

Salmonella is not the only thing to worry about with chickens. In 1997 Nichols Fox reported that "if there were a contest for the most contaminated product Americans can bring into their kitchens, poultry would win hands down. The USDA studies suggest that virtually every chicken is potentially infected with pathogens. If chicken were tap water, the supply would be cut off. Instead, chickens with heavy *E. coli* counts still receive the USDA's blessing of wholesomeness and safety."[26] That same year the Food Safety and Inspection Service (FSIS) finally made it unacceptable for feces to be visible on the skin of a chicken prior to its being immersed in chilling tanks where it might contaminate other birds. The lack of visible feces does not mean that feces and pathogens are not present. Indeed, a certain amount of fecal and pathogenic contamination is permitted in and on the chicken we consume. Ground chicken products may contain skin (as well as lungs and sex glands), which seems rather unwise, considering that skin is most vulnerable to contamination, and once it is ground into the meat there is no way to wash the contamination away. Also under current USDA rules, the added water weight permitted in chicken processing may contain chlorine. Finally, irradiation has been approved for use on chicken, although as this book goes to press the chicken industry is not utilizing the technology.

WHERE'S THE (SAFE) BEEF?

Imagine for a moment that during your lunch break, you and a colleague go to your favorite local restaurant and order cheeseburgers. Later that evening or early the next morning you both feel the beginnings of what you think is a little stomach "bug." You are in excellent health and just assume you have a bad case of stomach flu. Your friend has a weakened immune system and within days is rushed to the local emergency room. You are sick for a couple of days, but recover. Your friend has a long stay in the hospital and continues to be plagued with stomach problems for many months afterward. In either case, you both were probably exposed to *E. coli* and experienced varying degrees of food poisoning. It is true that illness due to *E. coli* has become more prevalent, but it is also more widely recognized than ever before—primarily due to the 1993 Jack-in-the-Box incident during which 700 people from four states were sickened with *E. coli* before the outbreak was contained. Four children died that year before action was taken.

The number of government agencies working diligently, yet somehow failing, to guarantee a safe food supply is astonishing. The USDA kept watch over the cattle as they were raised and the EPA inspected the water the animals drank. Both the USDA and FDA assured safe conditions at the slaughterhouse while the Department of Health and Human Services as well as state, city, and county health inspectors took care to guarantee sanitary restaurants staffed with trained food handlers and servers. The Department of Transportation, the state Department of Agriculture, and as many as eight other agencies could have been involved. As you might imagine, the red tape can be overwhelming. Even during a case as extreme as the one mentioned above, the government had a difficult time taking action: "When informed by the Washington State Department of Health of the connection between the Jack-in-the-Box outbreak and USDA-inspected meat, Jill Hollingsworth, assistant administrator of the Food Safety and Inspection Service (FSIS), told the health department's Charles Bartleson, 'We will take no action because this meat does not violate USDA standards.' To which Bartleson replied, 'I thought you guys were in the public health business.' "[27] The same thing that happened at the Jack-in-the-Box restaurant could happen in any city or town in America today.

Research is being done to reduce the risk of people getting sick from *E. coli*–tainted beef. In fact, a 1998 Associated Press article reported that cattle ranchers should "change what cattle eat for a few days before they are slaughtered. Feeding cows grain, as most farmers do to fatten them up, encourages the growth of *E. coli* bacteria that are strong enough to sicken humans, according to new Agriculture Department studies conducted at Cornell University."[28] Unfortunately, it is not that simple. As much influence as public pressure has on federal agencies to clean up our meat supplies, it often does not even begin to tip the scales when compared with the influence exerted by the meat industry and its lobbyists. David Korten, in *When Corporations Rule the World*, characterizes lobbying as "Washington's major growth industry which consists of the for profit public relations firms and business sponsored policy institutes engaged in producing facts, opinion pieces, expert analyses, opinion polls, and direct mail and telephone solicitation to create 'citizen' advocacy and public image building campaigns on demand for corporate clients."[29] The term "democracy for hire" has been used to describe these businesses, which are credited with providing us with a full 40 percent of the news we receive every day. Most large companies employ lobbyists, and the cattle industry is no exception. The latest figures show that $400,000 has been spent in one year by the National Cattlemen's Beef Association to help persuade the public and ultimately the government representatives that what is best for beef is best for the country. This time-honored tradition has its roots in the First Amendment of the U.S. Constitution, which specifies the right of the people to petition their government. It is a ubiquitous part of everyday life on Capitol Hill, and every segment of the agriculture industry is well represented. Lobbyists operate on the state, local, and national levels. Many lobbies even maintain offices in state capitals as well as in Washington, D.C. They can and do have a profound impact on the ways in which laws and regulations are made and enforced.

In 1999 the Clinton administration acknowledged these issues and advocated changing the Dietary Guidelines for Americans to include food safety concerns. The guidelines, which are reissued every five years, for the year 2000 may contain food safety information for the first time in their history. As the Associated Press reported, " 'I believe the time has come to include food safety in the guidelines,' Catherine Woteki, the Agriculture Department's undersecretary for food safety stated, 'the Government has an obligation to help people protect themselves.' " Woteki explained that even though the guidelines have never included this type of information, "times and dangers have changed since the early 1980s when the guidelines began.... Food safety is a critical factor in any discussion about diet and long-term health and new pathogens have emerged just in the last decade."[30] The question is whether lobbying efforts will forestall any action being taken by the government.

In the late 1990s, as a response to the confusion and misdirected intentions of the government agencies involved in the Washington State (Jack-in-the-Box) *E. coli* outbreak, a separate undersecretary for food safety was appointed. On the heels of that appointment came the enactment of a "zero tolerance program" for fecal matter on raw beef carcasses. Like myself, you might be surprised to learn that there was ever a time during which visible fecal contamination was permitted. It was actually processed, packaged, and sold to unsuspecting consumers with regularity before the Jack-in-the-Box incident. In 1996 the Clinton administration, also in response to the Washington State outbreak, published the final Hazard Analysis Critical Control Point (HACCP) regulation in an effort to assure food safety throughout a product's "life" from producer to consumer.

The HACCP system is a set of principles that strives to ensure safe food throughout all of the potentially hazardous areas in a food service operation or in a food service production plant. The system is a set of checks for

Lobbyist Spending 1997

(This list represents the highest-spending companies in each sector)

Industry	Lobbying expenditures	Campaign contributions	% Dem.	% Rep.
Chemical & Related Mfn.	**$29,975,745 (total)**	**$3,787,272**	**22%**	**78%**
• Arco Chemical Co.	• $1,160,000			
• BASF Corp.	• $1,470,552			
• Chemical Manufacturers Assn.	• $5,020,000			
• Clorox Co.	• $480,000			
• Dow Chemical	• $1,500,000			
• DuPont Co.	• $1,735,248			
• Eastman Chemical Co.	• $500,000			
• FMC Corp.	• $2,460,000			
• GAF Corp.	• $700,000			
• Kerr-McGee	• $410,049			
• Monsanto Co.	• $4,000,000			
• Olin Corp.	• $1,400,000			
• Proctor & Gamble	• $2,950,000			
Agricultural Services/ Products	**$10,852,887 (total)**	**$3,548,542**	**36%**	**64%**
• American Crop Protection Assn.	• $825,484			
• American Farm Bureau Federation	• $3,000,000			
• CF Industries	• $400,000			
• Dow Elanco	• $480,000			
• Farm Credit Council	• $420,000			
• Fertilizer Institute	• $646,000			
• IMC Global Inc.	• $300,000			
• National Council of Farmer Co-ops	• $639,678			
• Ralston Purina Co.	• $300,000			
Food Processing & Sales	**$9,401,655 (total)**	**$5,785,834**	**38%**	**62%**
• American Frozen Food Institute	• $260,000			
• ConAgra Inc.	• $350,000			
• Corn Refiners Assn.	• $440,000			
• Food Distributors Intl.	• $704,542			
• Frito-Lay Inc.	• $200,000			
• General Mills	• $781,000			
• Grocery Manufacturers of America	• $960,000			
• Kellogg Co.	• $335,681			
• Kraft General Foods	• $420,000			
• National Food Processors Assn.	• $559,685			
• Nestle USA Inc.	• $900,000			
• PepsiCo, Inc.	• $460,000			
• Pillsbury Co.	• $1,026,337			
Crop Production & Basic Processing	**$7,625,143 (total)**	**$5,233,232**	**41%**	**59%**
• American Peanut Product Manufacturers	• $240,000			
• American Rice Inc.	• $120,000			
• American Sugar Cane League	• $105,000			
• California & Hawaiian Sugar	• $120,000			
• Cargill Inc.	• $619,997			
• CBI Sugar Group	• $100,000			
• Chiquita Brands International	• $373,943			
• Domino Sugar Corp.	• $700,000			
• Flo-Sun Inc.	• $109,000			
• Georgia Commodity Commission for Peanuts	• $120,000			
• JG Boswell Co.	• $160,000			

Lobbyist Spending 1997 (continued)

Industry	Lobbying expenditures	Campaign contributions	% Dem.	% Rep.
Crop Production *(continued)*				
• National Corn Growers Assn.	• $280,000			
• National Cotton Council of America	• $333,818			
• National Oilseed Processors Assn.	• $120,000			
• Ocean Spray Cranberries Inc.	• $400,000			
• Philippines Sugar Regulatory Commission	• $120,000			
• Savannah Foods & Industries	• $110,000			
• Sugar Cane Growers Co-op of Florida	• $104,000			
• Sunkist Growers	• $118,000			
• U.S. Beet Sugar Assn.	• $440,000			
• U.S. Sugar Corp.	• $200,000			
• Western Growers Assn.	• $120,000			
Food & Beverage	$5,830,620 (total)	$4,805,948	29%	71%
• Burger King	• $160,000			
• Coca-Cola Co.	• $820,000			
• Hershey Foods	• $460,000			
• Mars Inc.	• $710,000			
• McDonald's Corp.	• $454,510			
• National Restaurant Assn.	• $840,000			
• National Soft Drink Assn.	• $926,000			
• Pizza Hut	• $260,000			
Dairy	$1,451,744 (total)	$1,263,146	38%	62%
• Associated Milk Producers	• $120,000			
• International Dairy Foods Assn.	• $416,000			
• National Milk Producers Federation	• $230,744			
• New Zealand Dairy Board	• $180,000			
Livestock	$1,182,000 (total)	$1,400,867	28%	72%
• National Cattlemen's Beef Assn.	• $400,000			
• National Pork Producers Council	• $200,000			
Fisheries & Wildlife	$992,710 (total)	$94,070	39%	61%
• American Pelagic Fishing	• $180,000			
• American Sportfishing Assn.	• $160,000			
• Chilean Salmon Farmers Assn.	• $100,000			
• National Fisheries Institute	• $120,000			
Poultry & Eggs	$490,000 (total)	$624,139	29%	71%
• National Broiler Council	• $120,000			
• Tyson Foods	• $140,000			
• National Turkey Federation	• $80,000			
• Perdue Farms	• $60,000			
Misc. Agriculture	$429,000 (total)	$150,331	34%	65%
• National Council for Agricultural Employers	• $260,000			

temperature and contamination that follows a product from the producer through production and transportation to the consumer. In 1998 a rapid, easy-to-use test for detecting *E. coli* in food products was introduced by a USDA biochemist. The test, which works on ground beef, is between 10 and 100 times more sensitive than other *E. coli* detection methods, requires no special training, can be done with minimal equipment, and takes only six to eight hours to confirm. Also in 1998 Health and Human Services Secretary Donna E. Shalala showcased PulseNet, a national computer network of public health laboratories that is capable of pinpointing food-borne illness outbreaks up to five times faster than by traditional methods. At the time of the 1993 Jack-in-the-Box incident, it took two weeks for the source of the outbreak to become known. If PulseNet works the way it should, it will take as little as forty-eight hours to identify the origin of an outbreak.

IS PORK BEING "PORKED"?

As 1998 drew to a close, the hog production industry faced a crisis. Recently there has been tremendous debate surrounding the hazards and methodology of industrial hog farming. As I described earlier, the odors that emanate from manure lagoons can debilitate humans, and the gaseous pollution produced from the same waste may even be deadly. Large-scale industrial farms, some housing as many as 330,000 sows, are forcing many small "family" pig farms out of business because they can produce pork much more cheaply than the small farmer can. For example, in December 1998 wholesale pork prices dropped to a low of 10¢ per pound—the lowest in thirty-four years. (Interestingly, retail prices remained the same and were down only a few cents from the previous year.)[31] With wholesale prices like those, small-scale pig farmers cannot afford to stay in business. Overproduction by industrial pig farms is primarily to blame for the crisis, because even if consumer demand for pork continues to slow, the mega-pork factories are too large to consider cutting back.

The National Pork Council continues to make frantic pleas to the government for assistance in stabilizing the market. One of the ways in which the government helps the pork industry is through subsidies. Agricultural subsidies have been part of the industry since the early 1900s, and they strive to help stabilize prices to ensure farmers a living wage as well as lower the cost of food to the consumer. They do this in a number of ways. A subsidy might pay a farmer more for his crop than the market will bear, it may pay a farmer to leave his land fallow, or it may even pay a farmer to farm more sustainably. No matter how a subsidy is distributed, it is the taxpayer that foots the bill. Subsidies are granted according to a farm's potential market value, which may be decided based on the farm's acreage, the number of animals it includes, or the potential market value of its livestock. As a result, factory farms are most likely to receive the largest percentage of our tax dollars. When subsidies are given to hog factory farms, both the tax-payers and the small farmers lose out. Taxpayers lose because they pay twice—once in the form of a subsidy, and then again at the checkout counter because subsidies often keep retail prices high. Small farmers lose because if they are farming more sustainably than factory farms, they receive a smaller share of the subsidized dollar and may go out of business because wholesale prices are low. Right now, taxpayers have no way of directing subsidies away from corporate factory farms, because many of the laws are created to favor corporate agriculture and subsidies are distributed according to farm size.

In December 1998 Secretary of Agriculture Dan Glickman announced that the government would purchase $50 million of pork to prop falling prices. This action might seem like a direct result of overproduction and price supports, but a *Time* magazine special report takes a second look. Seaboard Corporation is a multinational corporate factory pork farm that reports annual sales of over $1.8 billion and was the recipient of $100 million in subsidies in the late 1990s. The company's hog operation slaughters more than eight pigs a minute, or over 4 million a year, and there are plans to double that figure in the near future. Seaboard produces an enormous volume of waste. Pigs dead from disease and overcrowding collect at a rate of approximately 48 per hour, for an annual total of over 420,000. The *Time* article makes it clear that large companies with billions of dollars in revenues are often awarded subsidies not necessarily based on need, but on their ability to lobby governmental agencies based on a perceived need. It is not uncommon for these companies to have a negative impact on the environment, the animals they raise, and the people living near their facilities. In fact, the article points out that manure lagoons like those described in chapter 3 are located only about 75 feet from the Ogallala aquifer, which provides drinking water and irrigation to local inhabitants. Seaboard in particular took many of its subsidies, collected enormous profits, and closed up shop, leaving the town with the pollution, cleanup costs, and a severe shortage of jobs. Our tax dollars are paying for atrocities like this. And yet more help is on the way for the pork industry.[32]

THE "FRESH" FISH IN OUR DIET

In 1997 I spent a morning on the Boston fishing piers. It was an eye-opening experience for me. In my imagination there had always been something romantic about the fishing industry. I pictured swarthy fishermen standing on freshly scrubbed decks and small, well-worn boats bobbing contentedly along on calm seas, surrounded by a small flock of opportunistic seagulls. The reality was a far cry from my fantasy. What I saw was whole fish being dragged carelessly along cigarette-butt-littered concrete piers, fish cutters wiping their knives on bloody, dirty rags between jobs, and rotten fish waiting to be picked up for resale. There was an appalling disregard for sanitation. Naturally, many fish companies in Boston and all over the country have the highest regard for their products and their customers, but that trip to the Boston fish market made a deep impression on me.

The fishing industry, which has only recently become regulated, is under the control of both the FDA and the National Fisheries Institute (NFI). The FDA allows polychlorinated biphenyls (PCBs), dioxins, and DDT concentrations in fish. PCBs are allowed up to 2 parts per million, DDT (long outlawed yet still in the environment) 5 parts per million, and arsenic between 76 and 86 parts per million. Clams and oysters may contain an accumulation of lead between 1.5 and 1.7 parts per million, and governmental standards of acceptable mercury levels in swordfish have been raised to 1 part per million because too few fish could be found with lower levels.[33] PCBs are allowed in seafood at 2 parts per million, but dioxins and pesticides are often found in much higher concentrations in mollusks, freshwater fish, and saltwater species living close to shore.

Regulation of the seafood industry is behind the meat and poultry industries. However, the FDA's dramatic requirement that seafood processors and distributors utilize the HACCP system seemed to be a turning point for the industry in 1998. Thousands of company owners, plant managers, quality control personnel, and others put aside their daily tasks, attended training courses, mapped out how seafood should move through their facilities, and compiled in-depth HACCP plans for their operations. Product specifications, sanitation plans, and new importer-verification systems were also established. As the program's first anniversary approaches, *Seafood Business* describes its success as dubious: "Seventy percent of the 4,100 suppliers mandated for review by the FDA under HACCP regulations failed their first inspection." The article goes on to say that the FDA sent letters to all those who failed to comply and hoped that they would work with the FDA to rectify the situation. Thus, in spite of these new rules, the industry has a long way to go before use of HACCP makes a clear positive difference.[34]

SUSTAINABLE GOVERNMENT PROGRAMS

Prior to the 1980s, the USDA had done very little research on organic farming and other low-input production systems. Conventional agriculture benefited from most of the money spent on research, and virtually no attention was paid to alternative farming methods. The agricultural industry assumed the way to success was through better technology and enhanced efficiency. In 1979 the USDA's Science and Education Administration (SEA) conducted a study of organic farming in the United States to ascertain the benefits of addressing economic, energy, environmental, health, and farm structure issues through low-input and organic production. The result of the SEA study, published by the USDA as the *Report and Recommendations on Organic Farming* (1980), was unenthusiastically greeted by mainstream agriculture, which had hoped its methods would be hailed as the future of farming. In fact they were waiting for a replay of an earlier statement credited to former secretary of agriculture, Earl Butz: "We can go back to organic agriculture in this country if we must—we know how to do it. However, before we move in that direction, someone must decide which 50 million of our people will starve!"[35] In the face of the USDA's own report, Earl Butz was effectively advocating the continued commercialization of our nation's agriculture.

As much as proponents of sustainable farming practices hoped for support from their government, it would be many more years before the concept of agriculture free of chemical inputs and industrial farming practices would gain respect. By the 1980s many in the government and the agricultural sector were beginning to understand the potential pitfalls of agribusiness as a future agricultural paradigm. The 1985 Farm Bill, named the Food Security Act, was the first bill to propose programs aimed at reducing the adverse effects of conventional agriculture. The bill established a Conservation Reserve Program to reverse the erosion of farmland by taking corn, soybeans, cotton, and other highly erosive row crops out of production. It even went so far as to tell farmers that they would lose their eligibility for federal farm program benefits if they did not work to curb soil erosion on their land. Even though the title, "Agricultural Productivity Research," was carefully chosen to avoid terms such as *sustainable*, it was the forebear of a program called LISA (Low-Input Sustainable Agriculture) that strongly advocated sustainable farming practices.

The USDA described LISA in a 1988 brochure as a program that "helps keep farmers profitable by improving management skills and reducing the need for chemicals and other purchased inputs. It helps sustain natural resources by reducing soil erosion and groundwater pollution and by protecting wildlife."[36] It goes on to say that sustainable agriculture is a management strategy, the goals of which are to reduce input costs, minimize environmental damage, and provide production and profit over time. Established in 1985 as part of the Food Security Act, LISA first received funding in 1988. Its name was later changed to the Sustainable Agriculture Research and Education (SARE) program. SARE has funded over 1,200 projects through research and education, professional development, and producer grants. Total congressional funding for these programs in 1997 was over $11 million.[37] SARE also funded segments of Vermont's organic dairy industry and the Mothers & Others' CORE Values Northeast program, the longest-running SARE project, which encourages reduced chemical usage for northeastern apple farmers through implementation of IPM farming methods.

Vermont is mentioned frequently in discussions of sustainable agriculture programs, but many other states have instituted exemplary programs. In Texas, Jim Hightower, the former Texas commissioner of agriculture, made some impressive strides in supporting sustainable systems by supporting alternative forms of agriculture. One of Hightower's enduring efforts expanded the farmers' market program to include almost 2,500 new farmers. Another established "Texas Certified Organic," which assisted organic farmers in expanding their markets. A number of states have also instituted the "linked deposit" program, under which individual state governments deposit money in private banks at low interest rates, enabling banks to make low-interest loans available to farmers who intend to work toward alternative crop development—including farming with fewer or no chemicals. Iowa funds sustainable farming research and demonstration projects through a state tax on pesticides and nitrogen fertilizers. This form of taxation, often referred to as a "green tax," will be discussed at length in chapter 6; in a nutshell, the idea is to tax farming methods or purchases that increase pollution or environmental degradation. This tax money then goes to support the industry in ways that are sustainable and/or regenerative— in Iowa, funding research on low-input farming techniques. Two other states, Massachusetts and Maine, are also implementing creative projects. Massachusetts began the country's first Farmers' Market Program, which gives food coupons to low-income families to buy fresh fruits and vegetables at farmers' markets and inner-city farm stands. The program was so successful that the federal government now provides matching funds, and over twenty states use a combination of state, federal, and private funding to run their own coupon programs. The state of Maine passed two landmark food-labeling laws initiated by the Maine Organic Farmers and Gardeners Association (MOFGA)—one requiring retailers to label produce imported from countries that allow the use of pesticides banned in the United States, and another requiring labeling of any produce that has been treated with postharvest pesticides.[38] It is interesting to me that individual states are ultimately more adept at creating and maintaining these types of programs than is the national government. One shining example of the national government's inability to bring projects to fruition is the National Organic Standards Program. As this book goes to press, it has been in the pipeline for nearly ten years, with no end in sight. Many of the obstacles have been the work of politicians.

Senators and congressmen take up both sides of the sustainability debate. On the one hand some argue that as a country we should work to reduce agricultural chemical use. However, commodity group organizations and chemical and fertilizer companies often sway officials into backing a position that puts productivity and financial viability ahead of environmental sustainability. Today it is apparent that these two sides will never see eye to eye, but the gap is slowly being bridged between conventional and sustainable farming practices. In 1990 the Integrated Farm Management

Program Option was designed to allow farmers to receive commodity price supports for acreage left fallow during a shift from monocropping to the more sustainable practice of crop rotation and diversification. This program, part of the 1990 Farm Bill, was one of the first to financially support agricultural practices that were moving away from conventional agriculture and toward sustainable agriculture—an important first step, as up until that time almost all financial government support was dedicated to conventional farming. Still, ten years later less than 1 percent of government subsidy farming research goes to sustainable farming practices.

The hazards presented by chemically based agriculture remain a source of disagreement between profit-driven corporations and concerned citizens and politicians. In 1992 the Supreme Court unanimously

> upheld a decision by the lower courts authorizing local towns and cities to enact and enforce their own laws concerning pesticides. The town of Casey, Wisconsin, had passed ordinances that were stricter than federal standards and that had been immediately challenged in court by the chemical companies and their lobbyists. When Casey finally prevailed in the Supreme Court, trade associations (including the National Pest Control Association and the Professional Lawn Care Association of America) joined forces against the town. A trade journal was quoted as saying "legislation is rearing its ugly head in your community." Such groups have had bills sponsored in the Senate and the House that would preempt the Supreme Court ruling, disallowing local legislation of toxic pesticides. When Missoula, Montana, tried to pass a referendum calling for tighter controls on local spraying of chemicals, trade associations, along with Ciba-Geigy and DuPont, spent over $50,000 to defeat it.[39]

For many of these chemicals companies, the potential profits are so high that the money spent on lobbyists and court battles seems insignificant. Small towns have much smaller war chests and can be easily defeated. This is a situation that is sorely in need of a remedy. Small communities where families live and work should have control over the health and well-being of their citizens.

GOOD INTENTIONS GONE AWRY

Probably the most often recognized agricultural subsidy program is the Farm Loan Program, through which direct loans are offered by the FSA to those unable to obtain private or commercial credit. These loans are primarily used to purchase land, equipment, and seeds, or may even be used as operating capital. Another effort, the Conservation Reserve Program, is striving to conserve up to 36.4 million acres of land in various parts of the country, which will lead to improved water quality and protected wildlife habitats. The Reserve Program pays farmers to take designated environmentally sensitive land out of production, allowing it to remain "wild." The Emergency Conservation Program similarly shares with farmers the cost of rehabilitating farmlands damaged by natural disaster, such as drought. These are just a few of the ways that farmers can be supported; the government also provides price subsidies, irrigation supports, and direct farm aid. Unfortunately, so many of our government's best intentions do go wrong—not because our officials are lacking in dedication or integrity, but because the system itself is flawed. I believe that, for the most part, our elected officials do indeed have our best interests at heart. However, we are in a time when special interests are well financed. Lobbyists have tremendous power, and the issues are now so complex that it is often truly impossible to know what the correct plan of action is. Elected officials are often just as confused about the issues as we are. Vice President Al Gore expresses it this way:

> The new strategic threats to the global environment are becoming increasingly apparent, but do we understand how and why we created them in the first place? If our relationship to the ecological system is no longer healthy, how did we make so many poor choices along the way? For part of the explanation, we must look to politics. Too often, politics and politicians have not served us well on environmental issues, but there is also a fundamental problem with the political system itself. Aside from its uninspired response to the environmental crisis, our political system itself has now been exploited, mishandled, and abused to the point that we are no longer making consistently intelligent choices about our course as a nation.[40]

We seem to be on an endless, insidious merry-go-round that promises no chance of an alternative future outlook. The time has come to examine our options. If the U.S. government would encourage more renewable and sustainable agricultural practices, our farmers would very likely produce less, but American citizens would be relieved of the burden surpluses have placed on our taxes and there would be a reduced need for tax-funded government subsidies.

THE POLITICS OF FOOD

> We used to store mountains of excess grain in silos throughout the Midwest and let it rot, while millions around the world died of starvation. It was easier to subsidize growing more corn than to create a system for feeding those who were hungry.
>
> —Al Gore, *Earth in the Balance: Ecology and the Human Spirit*

The government works to boost agricultural profitability through price stabilization and commodity purchasing programs. The USDA receives a portion of our tax dollars to purchase surpluses, which are then distributed to needy families, the elderly, school lunch programs, and hungry people throughout the world. In 1999 these purchases were budgeted at almost $90 million.[41] On the surface, this is a well-intentioned program that works moderately well. However, there are some drawbacks, most specifically with regard to public health. Often our commodity programs are heavily weighted toward dairy and pork products, which, it is widely known, should be eaten only in moderation. Many of America's poor have access to little or no health insurance. Our tax dollars will also later be used to pay the cost of health problems due to improper diet—a diet that is effectively dictated by government subsidy and commodity programs. These decisions get made by our elected officials, often with the "help" of the lobbyists for commodity groups such as the pork and dairy councils. For example, in the first quarter of 1999, as earlier mentioned, the USDA agreed to purchase large quantities of pork to help support hog farmers, and some of this meat will be utilized in commodity food programs.

In 1998 the government committed to buying 80 million bushels of surplus wheat, later distributed to overseas food subsidies, to support farmers struck by climatic disasters. This may seem a bit confusing, but it is possible for our country to have an excess of stockpiled wheat while individual farmers are battling a bad season. If stockpiles overflow, prices are low, and when farmers have little grain to bring to a depressed market, their livelihood is at stake. This is when the government steps in to help farmers stay afloat. In a July 1998 weekly radio address President Clinton said, "Our farmers face a difficult and dangerous moment; many farm families have been pushed off their land, and many more could suffer the same fate unless our nation revives its commitment to helping farmers weather hard times."[42] This type of story, so often in the news, brings forth numerous other issues that have little to do with the moral satisfaction we may feel in spending our tax dollars to help small farmers. What we do here, in the United States, has a direct impact on the rest of the world. Government subsidies, the agriculture industry, the worldwide population explosion, and the stability of world peace are all woven together in a fragile fabric.

The United States is one of the largest agricultural exporters in the world, and through governmental programs our farmers are often called upon to provide food for starving nations. Using tax dollars, the government purchases goods for export from farmers at fair market value and ships them to other nations. However, if the farmers have produced a surplus, causing prices to fall, our tax dollars may once again be called upon to purchase additional product—not necessarily because we need more food for export, but to lower market supply in an effort to raise prices. These purchases will lower reserves and also raise prices, making it more difficult for us to buy food for the world's hungry in the future. Today, most of the farmers from whom the grain is purchased by the government are planting hybrid or bio-engineered seeds, which produce high-yielding plants. They are also utilizing the highest level of mechanized equipment, fertilizers, pesticides, insecticides, and herbicides—all in an effort to increase their production. In *State of the World 1998*, Lester Brown takes up these issues:

> The slower rise in world grainland productivity during the 1990s may mark the transition from a half-century dominated by food surpluses to a future that will be dominated by food scarcity, a time when growth in supply will lag behind the growth in market demand. In addition to the constraints on food consumption imposed by low incomes, rising prices could further restrict food intake among the poor, expanding the number of those who are hungry. If the politics of surpluses is replaced by a politics of scarcity, the issue will not be access to markets by a small handful of exporting countries but access to supplies by the more than 100 countries that import grain. Will exporting countries be willing to guarantee access to their supplies even in times of soaring grain prices?[43]

We must encourage our government to help support low-input and sustainable farming research and practices, because eventually our local food supply becomes an integral part of the global economy. As consumers we can affect the rest of the world through the choices we make on a local level. The adage "Think globally—act locally" comes to mind. The more we choose to participate in our local food webs and the more we work to influence our elected officials to support local economies, the less dependent we will become on the global food market. We may have better quality food, but there will be less for us to share with other nations. Brown continues:

> Stocks that will provide at least 70 days of world consumption are needed for even a minimal level of food security. Without this, one poor harvest can lead to a sharp rise in grain prices. Whenever stocks fall below 60 days' worth, prices become highly volatile. With the margin of security so thin, grain prices fluctuate with each weather report. When carryover stocks of grain dropped to 56 days of consumption in 1973, for example, world grain prices doubled. When they reached the new low of 52 days of consumption in 1996, the world price of wheat and corn—the two leading grains in terms of quantity produced—again more than doubled. A doubling of grain prices, were it to occur presently, would impoverish more people in a shorter period of time than any event in history. Instead of reducing the number of malnourished people from 800 million to 400 million by 2010, the goal adopted at the World Food Summit in Rome in late 1996, the ranks of the hungry would mushroom, dashing confidence in the capacity of governments to deal with this most pressing issue.[44]

We in America are privileged to enjoy an extremely prosperous and varied food supply, and many of us believe that world poverty has little to do with us. That is simply not true. If other countries starve and there is little or no intervention by wealthier nations, the global economy will begin to falter. With it, our own will suffer.

THE POLITICS OF FOOD AND POWER

Make no mistake about it, a government's power is often tied to its access to or control of food. In warring countries like Somalia and Kosovo, food is power. Contemporary sanctions imposed on countries like Cuba and Iraq have forced a shift in power. A nation's ability to control the food supply dictates its role in what becomes the politics of the global economy. One method by which countries maintain power over their own food supplies and protect their farmers is through trade agreements. All countries enforce trade barriers or agreements in an effort to command the highest prices for their producers. Every nation wants other countries to arrange reciprocal trade agreements.

Corporations from the United States as well as other economically powerful nations often use these "open door" policies to their advantage by manufacturing or growing their products in emerging nations where labor is cheap and taxation or other laws may be beneficial to the bottom line. During the latter part of the twentieth century, U.S. companies preferred in many cases to have their goods manufactured in China, Taiwan, Indonesia, or Mexico. In the agricultural industry this is referred to as biocolonialism, or a system by which a wealthy nation produces its high-cost food items in other countries. The reasons for this phenomenon vary, but frequently include lax chemical usage laws, inexpensive labor, lack of protective labor laws, and an absence of environmental degradation legislation.

The General Agreement on Tariffs and Trade (GATT) treaty and international trade organization has been in existence since 1948. The focus of much debate in the mid-1990s, it became a key campaign issue during the 1992 presidential elections as it related to the North American Free Trade Agreement (NAFTA). NAFTA, which called for the gradual removal of tariffs and other trade barriers on most goods produced and sold in North America, finally became effective in Canada, Mexico, and the United States on January 1, 1994.

GATT members worked to minimize tariffs, quotas, preferential trade agreements between countries, and other barriers to international trade. In 1994 GATT members and seven other nations signed a trade pact that would eventually cut tariffs and reduce or eliminate other obstacles to trade. Opening trade with other countries, and most especially with our northern and southern neighbors, initially seemed like a great idea, but there have actually been a multitude of complications involving money, agriculture, and the environment since NAFTA went into effect.

The following policy statement, which put profit motivation before the health and well-being of the population, illustrates the pitfalls: "Clayton Yeutter, in his capacity as U.S. Secretary of Agriculture under George Bush, stated publicly that one of his main goals was to use GATT to overturn strict local and state food safety regulations. He rationalized, 'If the rest of the world can agree on what the standard ought to be on a given product, maybe the US or EC will have to admit that they are wrong when their standards differ.' "[45] In other words, legislation favoring IPM or other low-input methods enacted by a state or town could be overturned by the federal government simply because the majority of countries involved in that segment of the food chain agree on lower standards of health and safety. This lemminglike approach is dangerous. The standards used by GATT, explains a Greenpeace study, "allow levels for at least eight widely used pesticides higher than current U.S. standards by as much as a factor of twenty-five, [and] the standards allow DDT residues up to fifty times those permitted under U.S. law."[46] The rules that govern agricultural standards for GATT are worked out for the World Trade Organization (WTO) by an organization known as the Codex Alimentarius Commission (Codex). Codex is an intergovernmental body established in 1963 and run jointly by the UN Food and Agriculture Organization (FAO) and the World Health Organization (WHO) to establish international standards on pesticide residues, additives, veterinary drug residues, and labeling. Delegations organized by the government and sent to Codex meetings often include nongovernmental representatives. Between 1989 and 1991, "one hundred forty of the world's largest multinational food and agrochemical companies participated and of a total of 2,587 individual participants, only twenty-six came from public interest groups."[47] That fact alone substantiates the idea that it is large business, not consumers or small farmers, that have a primary interest in securing lax trade regulations

The situation becomes even more difficult when we consider the laws of other countries in our decision-making processes. The importation of agricultural products accounts for $32 billion of food, equal to two and a half million shipments of food coming into the United States on a yearly basis, most of which is fruit and vegetables.[48] As mentioned earlier, although the U.S. government has outlawed certain agricultural chemicals, such as DDT, U.S. production of the chemicals does not necessarily stop. When the costs outweigh the gains for large corporations (in this case the cost of doing business in the United States would lead to zero profit), they often begin exporting their products to countries with less stringent regulations. Because so much food is produced for the United States in other countries, many of which allow the use of chemicals banned from use on American farms, what we end up with is a "circle of poison." The fact that DDT is banned in the

United States has virtually no impact; we are still consuming it on foods produced elsewhere.

Mothers & Others for a Livable Planet has been involved in issues surrounding pesticide usage for many years. In the organization's book *The Green Food Shopper: An Activist's Guide to Changing the Food System*, they explain the problem in detail:

> What lures agribusiness to grow food for the U.S. and Canada south—aside from rock-bottom wages and off-season sunshine—is freedom from pesticide regulation. Growers in Mexico, for example, regularly use at least six pesticides that are illegal in the United States. More often than not, these same pesticides come from the United States. According to a study last year by the Foundation for Advancements in Science and Education, from 1992 to 1994 the United States exported at least 344 million pounds of hazardous pesticides, of which at least 25 million pounds were forbidden in this country.[49]

The foundation then reported that "U.S. exports of restricted and severely restricted pesticides rose 33 percent from 1992 to 1996; of those, six pesticides considered 'extremely hazardous' by the World Health Organization skyrocketed more than 800 percent. Reported exports of pesticides banned in the U.S. remained steady, averaging around 6 million pounds a year."[50]

The government has long understood this to be a problem, and in 1991 the Circle of Poison Prevention Act was introduced in both houses of Congress, outlawing the export of pesticides that are either banned or unlicensed for use in this country. "In the blind pursuit of corporate profits," said Senator Patrick Leahy of Vermont, a sponsor of the bill, "U.S. chemical company giants dumped their poison overseas and devastated the lives of thousands of unsuspecting and innocent people."[51] Senator Leahy's referral to "unsuspecting and innocent people" is not accidental—all too frequently workers in other countries have no warning that the chemicals they are using are hazardous. Most of the time the pesticides are not even marked with warning labels. Even if warning labels are present or are posted for employees to read, they are often in English and cannot be understood by the workers. To complicate the issue, many countries allow children to work on farms. Not only do children have little or no understanding of the potential hazards presented by the chemicals, their small bodies are much more susceptible to the detrimental effects of pesticides and fertilizers.

Public Citizen, a consumer rights group, states that "while imports of Mexican fruit increased 35 percent and vegetables by 52 percent, . . . inspections of food for illegal pesticides fell by 5 percent between 1993 and 1995."[52] In fact, Mexico produces the majority of our imported off-season produce—97 percent of the tomatoes, 93 percent of the cucumbers, 95 percent of the squash, 99 percent of the eggplant, and 85 percent of the strawberries.[53]

Public Citizen goes on to say that "the most recent data indicates that imported food is more than three times more likely to be contaminated with illegal pesticides than domestically produced food." The Clinton administration promised, when NAFTA first went into effect, that if imports from Mexico increased, the FDA "would adjust the import program devoted to inspections accordingly."[54] As Public Citizen pointed out, this did not immediately come to pass.

In 1997 President Clinton enacted an "initiative to upgrade domestic food safety standards and to ensure that fruits and vegetables coming from overseas are as safe as those produced in the United States." The act requires the FDA to "halt imports of fruits, vegetables, and other food products produced in countries that do not meet U.S. food safety requirements."[55] The USDA worked with HHS to develop guidelines for these requirements, while at the same time the Department of Agriculture was set to the task of monitoring meat and poultry imports. It is a gargantuan undertaking, rife with problems. We import nearly 600 different types of food from nearly 150 different nations, totaling over 30 billion tons of product. Over 6 billion pounds of fruit, 8 billion pounds of vegetables, and 3 billion pounds of meat and poultry are imported into the country annually. Yet the FDA tests only 8,000 imported fruit and vegetable samples yearly, which means that on average roughly one residue test is performed for every 2 million pounds of food. FDA inspections decreased from 21,000 in 1981 to 5,000 in 1996, and to compound the problem, as of 1997 less than 700 FDA inspectors were responsible for twice as many imports as five years before. In fact, in 1997 there were less than a hundred import inspectors, each responsible for over 600,000 pounds of meat and poultry each week.

THE POLITICS OF POVERTY AND HUNGER

The U.S. government makes feeding women, children, and the poor a priority. Government food programs utilize a considerable amount of our tax dollars; in fact, a combined $40 billion is spent annually by the four largest assistance agencies. Most supplemental food programs are managed by the USDA and the Food and Nutrition Services (a subset of the USDA), whose mission is to "reduce hunger and food insecurity in partnership with cooperating organizations by providing children and needy families access to food, a healthful diet, and nutrition education in a manner that supports American agriculture and inspires public confidence."[56] The two most prominent poverty and nutrition programs are the Food Stamp Program and the Supplemental Food Program for Women, Infants and Children (WIC). In 1999 the Food Stamp Program helped put food on the tables of 8.3 million households and 19 million individuals every day, for a total of over $21.2 billion in benefits.[57] In spite of the government's dedicated efforts only about 60 percent of those who are eligible actually receive aid; the average grant amounts to a slim $72 per person per month, or 75¢ per person per meal. In the same year the WIC program fed over 7.4 million people, including 3.8 million children, 1.9 million infants, and 1.7 million women, for a total cost of $3.7 billion. That amount included $12 million for the WIC Farmers' Market Nutrition Program, which provides program recipients with coupons to be used only toward the purchase of fresh fruit and produce from farmers' markets.[58]

Children's supplemental meals programs have been a key component of the government's assistance package for a number of decades. In 1997 the National School Lunch Program fed nearly 26 million students each day in over 95,000 schools and residential child care institutions, for a total cost of over $5.2 billion. Additionally, state school lunch programs are entitled to choose from a list of over sixty different foods from the USDA commodity program. Unfortunately, commodity program products, which include items such as canned beef, chicken, and pork as well as salmon "nuggets," are not among the healthiest food choices. An extension of the lunch program is the School Breakfast Program, which fed an average of 7 million students each day in 1997. Available in approximately 69,000 schools, it provided a total of nearly 1.2 billion meals during that same year.[59] To accommodate the need for supplemental meals in the summer, the FNS implemented the Summer Food Service Program, which in 1997 fed more than 2.3 million children per day at over 28,000 sites.[60]

No matter how productive these governmental programs may seem, none is without need of improvement. Increased flexibility within existing systems could be of great benefit to many of the current supplemental programs. I spoke with Ellen Haas, undersecretary of food nutrition and consumer services from 1993 to 1997. While she worked on several programs during her tenure there, the Food Stamp Program and its nutrition component were of particular importance to her. One of Ellen's goals for the Food Stamp Program was to create a core educational system that would teach basic nutrition, enabling aid recipients to make good, healthy food choices. Unfortunately, her plans to change the program were met with intense resistance. She explained,

> You have bureaucracies that have worked in one way for many years, and change is the most frightening thing. I had 1,800 people who worked for me, and in the school lunch program there were several hundred who only cared about counting the number of children, and making sure everything fit into its little box. When someone comes in and talks about how healthy the food should be, how tasteful it should be, and that children should be educated, some of those people feel threatened because that is not the job that they were brought there to do.[61]

USDA Foods Expected to Be Available in School—Fiscal Year 2000*

Schools and Institutional Programs

I think most would agree that one of our most important assets is our children, and they need to be well fed and educatedfor them to rise to their highest potential. I find it extremely disheartening that most school lunch programs are allocated less than one dollar to feed the next generation their lunch five days a week. Compound that with some of the foods in the following list and the situation becomes truly horrifying.

Below is a list of the commodities that the USDA expects to be available to State Distributing Agencies in School Year/Fiscal Year 2000 for schools participating in the National School Lunch Program (NSLP) or the School Breakfast Program (SBP); and for institutions participating in the Child and Adult Care Food Program (CACFP), the Summer Food Service Program (SFSP), and the Nutrition Program for the Elderly (NPE).

SECTION 32 TYPE DONATED COMMODITIES
SCHOOLS/INSTITUTIONS
Group A Products
(Meat/Fish/Poultry/Fruits/Vegetables)

Meat/Fish
- Beef, Ground, Frozen—36 lb. carton
- Beef, Patties, Frozen: 100 percent, VPP & Lean—36 lb. carton
- Beef, Canned w/Natural Juices—24/29 oz. can (offshore only)
- Ham, Cooked, Water-added, Frozen—4/9 lb. hams per carton
- Ham, Roast, Frozen—32–40 lb. carton
- Meat, Luncheon, Canned—24/30 oz. can (offshore only)
- Pork, Canned w/Natural Juices—24/29 oz. can (offshore only)
- Pork, Ground, Frozen (Fine)—36 lb. carton
- Pork, Sausage Bulk, Frozen—36 lb. carton
- Pork, Sausage Mild VPP, Frozen: 1.5 & 3 oz. patties and bulk
- Pork, Sausage Patties, Frozen: 1.5 & 3 oz. patties—36 lb. carton
- Pork, Sausage Spicy VPP 1.5 & 3 oz. patties and bulk

Poultry/Egg Products
- Chicken, Cut-Up, Frozen—40 lb. carton
- Chicken, Cooked, Breaded, Frozen—30 lb. carton
- Chicken, Diced, Frozen—40 lb. carton
- Chicken, Canned Boned—12/50 oz. can
- Chicken, Drumsticks, Frozen—40 lb. carton
- Chicken, Thighs w/Backs, Frozen—40 lb. carton
- Chicken, Leg Quarters, Frozen—40 lb. carton
- Eggs, Frozen, Whole—6/5 lb., 30 lb. carton

- Egg Mix—4/10 lb. bags
- Turkey, Burgers, Frozen—36 lb. carton
- Turkey, Ground, Frozen—4/10 lb. carton
- Turkey, Hams, Frozen—40 lb. carton
- Turkey, Roast, Frozen—32–48 lb. carton
- Turkey, Sausage, Chubs, Frozen—30 lb. carton
- Turkey, Whole, Frozen—30–60 lb. carton

Meat/Poultry/Fish Commodities for Processing
- Beef, Coarse Ground, Frozen—60 lb. carton
- Beef, Special Trim, Frozen—60 lb. carton
- Chicken, Bulk, Chilled—small or large
- Eggs, Liquid, Bulk—tankers
- Pork, Coarse, Frozen—60 lb. carton
- Turkey, Bulk, Chilled—90 lbs.

Meat/Poultry/Fish Available Under State Option Contract (SOC) Program
- Beef, Cooked Patties, Frozen: 100 percent & VPP—36 lb. carton
- Chicken Nuggets—30 lb. carton
- Chicken Patties—30 lb. carton
- Pork, Cooked Rib Patties, Frozen—36 lb. carton

Fruits/Vegetables (Canned, Dry, Frozen)
Fruits
- Applesauce, Canned—6/#10 cans
- Apple Slices, Canned—6/#10 cans
- Apple Slices, Frozen—30 lb. carton
- Cherries, Frozen—30 lb. carton, IQF (Individually Quick Frozen)—40 lb.
- Cherries, Water Pack, Canned—6/#10 cans
- Fruit Mixed, Canned—6/#10 cans
- Orange Juice, Concentrate, Frozen—32 oz. can, tankers & drums
- Peaches, Canned, Clingstone, Sliced/Diced—6/#10 cans

USDA Foods (continued)

- Peaches, Frozen—20 lb. case, 96/4 oz. cups
- Pears, Canned (halves, sliced, diced)—6/#10 cans
- Pineapple, Canned (tidbits, chunks, crushed)—6/#10 cans
- Plums, Canned, Whole—#10 cans

Vegetables
- Beans, Dry—25 lb. bag
- Beans, Dry, Canned—6/#10 cans
- Beans, Green, Canned—6/#10 cans
- Beans, Green, Frozen—30 lb. carton
- Beans, Refried, Canned—6/#10 cans
- Beans, Vegetarian, Canned—6/#10 cans
- Carrots, Canned—6/#10 cans
- Carrots, Frozen—30 lb. carton
- Corn, Cobbettes, Frozen—30 lb. carton
- Corn, Frozen—30 lb. carton
- Corn. Liquid, Canned—6/#10 cans
- Corn, Vacuum, Canned—6/75 oz. cans
- Peas, Canned—6/#10 cans
- Peas, Frozen—30 lb. carton
- Potato Rounds, Frozen—6/5 lb. package
- Potato Wedges, Frozen—6/5 lb. package
- Potatoes, Oven Type, Frozen—6/5 lb. package
- Salsa, Whole—55 gal. drum
- Spaghetti Sauce/Paste—6/#10 cans
- Sweet Potatoes, Frozen—6/5 lb. package
- Sweet Potatoes, Mashed, Whole—6/#10 cans
- Tomato Paste, Bulk—55 gal. drum
- Tomatoes, Canned: Sauce/Diced/Whole—6/#10 cans

Fresh Fruits and Vegetables
Fresh Fruits
- Apples, Fresh—37–40 lb. carton
- Grapefruits, Fresh—34–39 lb. carton
- Oranges, Fresh—34–39 lb. carton
- Pears, Fresh—45 lb. carton

Fresh Vegetables
- Potatoes, Fresh, White, Russet—50 lb. carton

SECTION 416 TYPE DONATED COMMODITIES—SCHOOLS/INSTITUTIONS

Group B Products (Grains/Cereals/Cheese/Milk/Oils/Peanut Products)
- Bulgur—50 lb. bag
- Cheese, Cheddar, Reduced Fat (white, yellow)—8/5 lb., 40 lb. block
- Cheese, Cheddar (white & yellow)—4/10 lb., 40 lb. block
- Cheese, Cheddar, Shredded (white, yellow)—8/5 lb.
- Cheese, Process (loaves, sliced)—6/5 lb.
- Cheese Cheddar, Reduced Fat, Shredded (white, yellow)—8/5 lb.
- Cheese, Mozzarella, Frozen (small, all)—3/20 lb. loaves; Unfrozen all
- Cheese, Mozzarella, Low Moisture Part Skim (LMPS), Frozen & Unfrozen—3/20 lb. loaves
- Cheese, Mozzarella, LMPS, Shredded, Frozen—15 lb. 30 lb. & 45 lb.
- Cheese, Mozzarella Lite, Frozen (small);
- Cheese, Mozzarella Lite, LMPS, Shredded, Frozen (small)—15 lb., 30 lb., & 45 lb.
- Cheese, Natural American, Barrel—500 lb.
- Cheese Blend, American/Skim Milk, Sliced—6/5 lb.
- Cornmeal—8/5 lb. bags, 4/10 lb. bags, 50 lb. bags
- Flour, all types—4/10 lb. bags, 8/5 lb. bags, 50 lb. bags, 100 lb. bulk bags
- Flour, Bakery Mix, Lowfat—6/5 lb. bag, 35 lb. carton
- Flour, Corn Masa—50 lb. bags
- Grits, Corn (white/yellow) —8/5 lb. bags, 4/10 lb. bags, 40 and 50 lb. bags
- Milk, Nonfat Dry—25 Kg. bags 2/
- Macaroni (elbow/spiral/rotini)—20 lb. carton
- Oats—50 lb. bags
- Oil, Vegetable—6/1 gal. bottles, bulk, 8/84 oz. bottles
- Oil, Vegetable Saturated Reduced-Fat—6/1 gal. bottle 1/
- Peanut Butter (smooth, chunky)—4 or 5 lb. cans
- Peanut Granules—6/#10 cans
- Peanuts, Roasted—6/#10 cans
- Rice, Milled—25 lb. bags, 50 lb. bags
- Rice, Parboiled Milled 25 lb. & 50 lb. bags
- Salad Dressing, Reduced-Calorie/Regular)—4/1 gal. bottle
- Shortening, Vegetable—12/3 lb. can, 50 lb. cube
- Shortening, Liquid Vegetable—6/1 gal. bottle
- Spaghetti—20 lb. carton
- Rice, Brown—25 lb. bags
- Rice, Milled—25 & 50 lb. bags
- Rice, Parboiled, US#1 Large Grain—25 & 50 lb. bags

*From the USDA website: www.fns.usda.gov.

Ellen believes that the Food Stamp Program should be more "a nutrition program and not a poverty program"; she concedes that "it does support incomes, but [it] also supports the health of the individuals who are poor." One program that aspires to be an educational tool, rather than one based on a system of handouts, is Operation Frontline. I cotaught an Operation Frontline course whose purpose it was to teach women in the WIC program healthier ways of eating, cooking, and shopping. Many of the people I met were uneducated about food choices, proper nutrition, and the amount of salt, additives, and chemicals in processed foods. I came away from my experience believing that all food supplement programs should include education, but when I spoke to Ellen, she told me that the WIC program was the only one of its kind.

As a member of the Welfare Reform Task Force, Haas endeavored to be heard on the issues surrounding the health of program participants but never felt as though she was reaching those who could make a difference. Many in positions of power believe that even when the government is handing out free food (in this case, in the form of food stamps), it has no right to dictate what types of food people should be eating. It is a "freedom of choice and civil rights issue," Haas explained. The WIC Farmers' Market Nutrition Program appears to be a strange exception to the rule. More than 20 million Americans receive food stamps; half of the recipients are children, and half of the families are single-parent households. Haas believes that with numbers like those, food education should be a number one priority—yet it never has been. Change in Congress is excruciatingly slow and frustrating, she said; "It is like moving a glacier."

During the time she worked on the Food Stamp Program, Haas did have some "small successes." She said, "If you go into a food stamp office it is one of the most depressing places in the world, and the walls are bare. We had a poster contest that turned out terrifically, with at-risk students utilizing nutrition messages, and they are now in offices all around the country." Another of Haas's successful projects was Team Nutrition, born out of Haas's drive to elevate the quality of the school lunch program. She believes the program, whose goal it is to bring food education to the classroom, is extremely important; many children are so completely detached from the sources of their food that they believe that it comes from a drive-up window or a microwave pouch. (In fact, during a recent interview on National Public Radio, Alice Waters noted that 85 percent of American children do not sit at a table for a meal on a regular basis.) Haas was eventually instrumental in coordinating a collaborative effort that includes the American Culinary Federation, but it all started with the school lunch program.

The fact that the school lunch program was dreadfully lacking in creativity, balance, and nutrition did not go a long way toward disabusing children of their misunderstandings about the sources of their food. Haas told me that she feels strongly that nutrition is a bridge between agriculture and health. Educating consumers, especially children, is a way of making that connection even stronger, and the school lunch program was sending the wrong message to kids all around the country. Through a nationwide study, Haas determined that school lunches were 25 percent higher in fat and 100 percent higher in sodium than recommended by the government's own USDA dietary guidelines. Additionally, less than 2 percent of the food being served consisted of fresh fruits and vegetables. The investment being made in school lunch at that time, per child, including milk, was under $1. It is easy to see why the food was so appallingly bad.

With the results of the study, Haas managed to pull together the beginnings of Team Nutrition in a small campaign known as Fresh Start, which aims to increase the amount of fresh foods in school lunches. Its resources quickly became overburdened with the challenge of meeting the needs of 96,000 schools nationwide. Even though program policies were being made in Washington, purchasing decisions continued to be made by schools at the local level. Each school had its own hurdles to cross, and Fresh Start was responsible for providing necessary assistance. Cost and delivery problems existed across the board. It was Haas who came up with a viable solution. The defense department, with troops all over the country and the world, is the largest government food purchaser. Haas solicited their involvement, which, for Fresh Start, lowered costs and increased nationwide delivery. All schools then had access to inexpensive fresh fruits and vegetables. This initiative led to the establishment of School Meals for Healthy Children, whose core goal was to increase the amount of fresh fruits and vegetables as well as grains in the school lunch program as a way of promoting good health. The next step in the process was the beginning of what actually became Team Nutrition. Ellen elicited the assistance of chefs around the country to implement an education campaign that began with the training of school food service workers around the country. In all, 500 chefs agreed to participate in the Great Nutrition Adventure, during which they went into the schools and worked side by side with school kitchen staff. When Haas left the Department of Agriculture, Team Nutrition was in 35,000 schools, and today the majority of schools utilize at least some of the materials. I can personally attest to the quality of the teaching tools generated by Team Nutrition. Among the materials I received were cooking videos, recipe books, sanitation education manuals, and tips for healthy, fresh cooking techniques. I asked Ellen whether any studies have been done to establish the long-term viability and success of programs like Team Nutrition, and her answer was disappointing: "In my last year, $2 million was spent on a study, which has still not come out." She explained that the delay is due to the fact that the results were "too good." For political reasons and because of some people with the "mentality of career bureaucrats," the results were sent back to be reviewed. She believes that when the results are finally released, they will describe wonderful successes.

One of the outgrowths of the Team Nutrition project is Team Nutrition Gardens in California, which incorporates Alice Waters's Edible Schoolyard program. During a competition sponsored by the American Culinary Federation (ACF) I was able to see Team Nutrition in action and was impressed with the program's ability to make a difference. The goal was to make healthy, nutritious, and flavorful school lunches within the program's dietary and monetary guidelines. Competitions were held at ACF National Conventions in 1994 and 1995, and a recipe book generated through the competitions was published in 1996. A number of the judges were school-age children, and all of the recipes showcased at the competition were distributed to kitchen workers in schools across the country.

The government truly touches everything we eat and drink. To this point we have discussed agricultural history, the sustainability of our food supplies, and even agribusiness. With a better understanding of where the government fits in the big picture, it is important to bring the story home—right to your kitchen table. The next chapter takes a look at what all of this has to do with our food choices, our health, and our daily lives.

▪ 5 ▪

We Are What We Eat

Tell me what you eat, and I will tell you what you are.
—Anthelme Brillat-Savarin, *The Physiology of Taste*

OUR DAILY BREAD

Food is integral to all our lives. From it we derive pleasure; with it we create celebrations and nurture one another. Holiday meals are spent with loved ones, chicken soup heals a cold, and on Valentine's Day chocolate is a gift of decadence for friends and lovers. These days, the pace of our lives has increased dramatically, and we have less time than ever for preparing and enjoying meals. The busiest among us may have holiday meals catered, and chicken soup is almost always available in a can. Instead of being cooked and served to us by loved ones, our food is now created in large, industrial kitchens by people we do not know. The goal of anonymous food handlers is not to nurture us but simply to feed us. There is little to no emotional connection between the feeder and the fed. Errors of judgment by today's food processors occur with greater frequency and are less likely to be noticed unless there is a reported outbreak. Food is produced on a mass scale, animals pumped full of hormones and other growth-promoting drugs are raised in close quarters, and virtually everything we eat is being shipped across long distances. Processed food products are preserved with any number of chemicals, enhanced with additives, and frequently contaminated with pesticides and herbicides. Time saved has come to be more valuable than our health. We eat a tremendous amount of processed and prepackaged foods. Salt, fat, and refined sugars are in high supply, and our diets and health are suffering. The results for many are cancer, heart disease, high blood pressure, diabetes, and obesity. Everything we consume becomes, as the French gastronome Brillat-Savarin implies, part of us. Our food provides us with nourishment. It sustains us. We literally are what we eat—good and bad.

AN IMPOVERISHED DIET AND THE EXPANDING AMERICAN WAISTLINE

Americans eat huge quantities of food. Currently the average American consumes 3,800 calories a day, a figure that represents a 15 percent increase over the past two decades. The USDA estimates that a 125-pound woman who exercises moderately needs 2,250 calories to maintain her weight. Likewise, a 175-pound moderately active man requires about 3,500 calories. Both are well below the average cited. Of those 3,800 calories, research indicates that we consume 150 to 250 percent more fat and 175 to 245 percent more protein than what the USDA recommends for a healthy diet. It should come as no surprise, then, that we are a nation of fat people. In fact, the U.S. National Center of Health Statistics estimates that 31 percent of women and 35 percent of men are overweight, while less optimistic independent researchers say that 54 percent of all adults in the United States are of unhealthy weight. This is particularly alarming since forty years ago only 5 percent of American adults were overweight. If the current trend continues, within a few generations virtually every American adult will be overweight. A large percentage of American children are already obese, and an incredible 40 percent are just plain fat. A study by the Institute of Medicine estimates that obesity costs the United States about $70 billion annually in direct health care expenses or in lost productivity.[1] Add that to the $50 billion spent on diet aids, and the total rises to $442 spent every year on every person in this country. That figure is equal to more than 30 percent of what each of us spends on food annually.

To be sure, a great deal of America's problem with obesity lies in simple overconsumption, but we are also eating the wrong kinds of foods. Beef, pork, lamb, and veal are animal proteins containing the highest percentage of fat, and they account for the majority of meat products consumed by Americans today. Fish and shellfish, with the lowest fat content, are least frequently ordered in restaurants or purchased in markets. In the United States we eat only about fifteen pounds of seafood per year. We eat three times as much chicken and five times as much beef. The steak I order from my supplier as a chef costs me $3 to $4 more per portion than salmon and $2 to $3 more than shrimp. Strangely, we prefer to eat animal proteins that are less healthful and more expensive than fish.

Animal fats are not singularly responsible for our dietary problems. In the past twenty years, Americans have increased consumption of vegetable fats and oils by 40 percent. By consuming huge amounts of fast food, fried food, and sugar, we are literally eating our species into decline. Martin Teitel elucidates in *Rain Forest in Your Kitchen*: "Fast food is nothing if not pervasive in the United States where 50 percent of us live within fifteen minutes of some chain outlet. Fast food is a wonderful and terrible thing. On a typical day, nearly half of all adult Americans patronize some type of food service establishment, reports the National Restaurant Association. And according to industry association projections, roughly 80 percent of all dollars spent on eating out will pass over the counters of fast-food establishments."[2] All of our food woes are not centered on fast food. A 1998

Newsweek article focused on American sugar consumption: "Millions of us start the day with dessert, whether it is frosted Pop-Tarts, marshmallow-laden cereal, chocolate yogurt, or a muffin the size of a small TV. Then we spend the next 14 hours chugging more sweets, from a midmorning danish to a vanilla shake at lunch to cookies after dinner. Meanwhile, soft drinks rise around us like floodwaters: we are swigging nearly 54 gallons a year per person."[3] The amount of sugar and other caloric sweeteners (such as honey, molasses, and corn syrups) we consume recently hit an all-time high. In 1997 the average American consumed 154 pounds of caloric sweeteners, an increase of 75 percent since 1909. Soft drinks are primarily responsible for this dramatic increase. In fact, today we guzzle over 110 percent more soda than we did twenty years ago. The National Center for Health reported in 1998 that 300,000 deaths annually are attributable to poor diet and inactivity. Nearly 70 percent of the diagnosed cases of cardiovascular disease are related to obesity.[4] Obesity increases the risk of kidney, colon, ovarian, endometrial, gall bladder, cervical, prostate, rectal, and breast cancers, and is a risk factor for stroke and adult-onset diabetes.

The highest rate of obesity and diabetes in the United States is among Native Americans and Hawaiians. In 1998, the *New Yorker* published an article on the "Pima Paradox," describing a particular group of Native Americans with the highest rate of diabetes of any ethnic group in the country. A recent development in Pima history, this propensity is believed to be directly attributable to diet and culture. Among adults aged thirty-five and over, the *New Yorker* reported,

> The rate of diabetes is fifty percent, eight times the national average, and a figure unmatched in medical history. It is not unheard of for adults to weigh five hundred pounds, for teen-agers to be suffering from diabetes, or for relatively young men and women to be already disabled by the disease—to be blind, to have lost a limb, to be confined to a wheelchair, or to be dependent on kidney dialysis. One hypothesis for the Pima's plight, favored by Eric Ravussin, of the National Institutes for Health (NIH), is that after generations of living in the desert the only Pima who survived famine and drought were those highly adept at storing fat in times of plenty. Under normal circumstances, this disposition was kept in check by the Pima's traditional diet: cholla-cactus buds, honey mesquite, povertyweed, and prickly pears from the desert floor; mule deer, white-winged dove, and black-tailed jackrabbit; squawfish from the Gila River; and wheat, squash, and beans grown in irrigated desert fields. By the end of the Second World War, however, the Pima had almost entirely left the land, and they began to eat like other Americans. Their traditional diet had been fifteen to twenty percent fat. Their new diet was closer to forty percent fat. Famine, which had long been a recurrent condition, gave way to permanent plenty, and so the Pima's "thrifty" genes, once an advantage, were now a liability.[5]

Similarly, Hawaiians whose dietary habits now include lots of junk food and fat show critically high percentages of diabetes and coronary disease. Hawaii's statistics on longevity may have earned it recognition as the "Health State," but a look beyond the superficial facts reveals that native Hawaiians are actually the unhealthiest people in the United States. In an excerpt from *The Hawaiian Paradox*, Terry Shintani reports that

> in terms of the causes, the paradox suggests that the destruction of the cultural ways and abandonment of the traditional diet of the native Hawaiians have resulted in staggering rates of death from chronic disease. . . . In ancient times, Hawaiians were believed to be a very healthy, robust race of people, and early accounts suggest that there was little heart disease, cancer, diabetes, or obesity. . . . In 1987 as a survey of the mortality of pure Hawaiians indicated staggering rates of mortality from chronic disease including heart disease (278 percent higher), cancer (226 percent higher), stroke (245 percent higher), and diabetes (an astounding 688 percent higher) . . . than the national average . . . the destruction of cultural ways and abandonment of traditional diet has been the primary cause of these high rates of disease among the native Hawaiians.[6]

Native Americans and Hawaiians, while displaying unusually high rates of illness, are not alone in continuing their poor dietary habits in spite of advice to the contrary. Most Americans believe a healthy diet is important for their long-term well-being but are quick to admit they do not eat right. In 1997, the *Christian Science Monitor* reported that "about 65 percent of Americans say it is important to eat fruits and vegetables, but only 40 percent consistently do. Only 10 percent of Americans eat a green or yellow vegetable daily."[7] Everyone knows that a good diet includes lots of fruits and vegetables—our mothers have been telling us that forever, right? So why are we not eating them? Why are half of all Americans taking dietary supplements to augment their diets? Laziness? Perhaps partially. While there are many theories around this particular issue, I believe that part of the reason is that while we listen to and absorb the age-old rhetoric on fruits and vegetables, there are not many of us who actually know *why* we should eat them. We also do not know which ones provide us with which nutrients.

It appears that not only are fruits and vegetables low in fat and high in vitamins and minerals, but they can actually help fight disease. Some of this information was discovered through research that attempted to understand the sudden changes in health among our indigenous peoples. It was found that Native Americans who returned to traditional diets and followed a regular exercise program began to show remarkable improvements in their health. Some who had been insulin-dependent for years began to live without insulin. Testimonial after testimonial revealed that there was a way to avert a seemingly hopeless situation; the answer had been reflected all along in the foods of their cultures. More and more studies are proving that the foods we are not eating could impact our health as much as the foods we are eating.

FIGHTING DISEASE WITH DIET

In the United States, more than 2 million people develop cancer each year and about 560,000 die of the disease. It is estimated that diet and obesity are responsible for about 30 percent of all cancer deaths while sedentary lifestyle is responsible for about 5 percent.

—*The Strang Cookbook for Cancer Prevention*

Published in 1998, *The Strang Cookbook for Cancer Prevention* makes a clear case that what we eat directly impacts our health. The authors believe that a healthy diet of fruits, vegetables, whole grains, and low-fat foods not only helps reduce the risks of obesity, diabetes, and heart disease, but can actually prevent cancer.[8] In our interview, author Laura Pensiero explained that "by fleshing out all the other confounding factors, experts have estimated that 1 to 2 percent of cancers are environmental, whereas 40 percent are related to diet, and most have to do with the foods we are not eating— like our very low intake of whole foods and fruits and vegetables. When they studied the Mediterranean diet in Crete, all the news had been on heart disease; this year some studies came out linking the Mediterranean diet with lower rates of cancer."[9] Laura believes that a full 80 percent of cancers can be prevented through healthful diet combined with exercise. A November issue of *Newsweek* reported in 1998 that researchers were finding that a wide variety of chemicals common to fruits and vegetables could slow and even stop the growth of tumors in laboratory animals. The article reports, "Today our knowledge of these compounds is exploding. And as scientists learn more about the chemistry of plants and other edibles, they grow ever more hopeful about sparing people from malignancy. . . . Eating the right foods is 'as specific to stopping cancer before it starts as wearing a seat belt is to lowering your risk of a fatal automobile accident.' "[10] Carotenoids, vitamins A, C, D, and E, selium, calcium, folic acid, and magnesium are all believed to work against cancer in the human body. Functional foods, also known as phytochemicals, are plant foods believed to provide health benefits beyond basic nutrition. They include broccoli, kale, cauliflower, and mustard greens and are currently the subject of much scientific research. Preliminary findings indicate that phytochemicals are also anticarcinogens. While this is not yet proven, research into the anticarcinogenic properties of food nutrients is ongoing, and the theories put forth by these researchers are gaining respect throughout the scientific community. In fact the National Cancer Institute has made potential cancer-fighting foods the focus of their research, in 1999 releasing a report that tomatoes and tomato products reduce the risk of cancer.[11]

Mitchell Gaynor, head of medical oncology at New York's Strang Cancer Prevention Center, believes that the future of cancer prevention is in food. He uses tomato sauce as an example: "There's lycopene, the pigment that gives tomatoes their blush. You can't get much lycopene from a raw tomato; it's too tightly bound by the fruit's proteins and fibers. But cooking frees it for absorption by the body, and dietary fat helps carry it into the bloodstream. If that's not reason enough to enjoy some tomato sauce with a dash of olive oil, consider this. In a 1995 study of 48,000 men, Harvard researchers found that those who ate 10 servings of tomato-rich foods every week cut their risk of prostate cancer by nearly half. Other studies suggest that lycopene may help ward off cancers of the breast, lung and digestive tract."[12]

That is the good news. The bad news is that new food composition tables posted on the USDA's website show a sharp decline in vitamins and minerals in American-grown foods since the last survey, which was conducted over twenty years ago (see Table 5.1). Broccoli, for instance, has lost 17.5 percent of its vitamin C and 53.8 percent of its calcium. Likewise, a random sampling of twelve common garden vegetables compiled by Alex Jack, editor of the Kushi Institute's *One Peaceful World*, showed an average decline in calcium of 26.5 percent, in iron of 36.5 percent, of vitamin A of 21.4 percent, and in vitamin C of 29.9 percent.[13] The loss of healthy soil with high concentrations of nutrients and vitamins in concert with agriculture's increasing reliance on chemical fertilizers is responsible for much of this decline.

Not only are fish and shellfish low in fat, but research has shown that their omega-3 fatty acids may also lower the risk of cancer and reduce the growth of tumors. In fact, the Japanese, whose per capita seafood intake is ten times ours, have a significantly lower cancer rate than Americans. Naturally, genetic factors play a role in the differing cancer rates. However, if we accept Laura Pensiero's premise that 80 percent of all cancers can be prevented through diet and exercise, it is safe to assume that the composition of the Japanese diet is significantly related to their low cancer rates.

Rebecca Goldberg, author of *Murky Waters: Environmental Effects of Aquaculture in the United States*, tells us that not only does fish consumption help reduce the risk of cancer, but "omega-3 fatty acids . . . can lower blood levels of triglyerides (fats) and reduce the frequency of blood clotting." She goes on to say, "Recent studies suggest that diets rich in fish oil can reduce the severity of heart attacks by affecting the heart's electrical mechanisms, and can have beneficial effects in some people suffering from diabetes or inflammatory or allergic diseases," such as rheumatoid arthritis.[14] Additionally, omega-3 fatty acids have been shown to help reduce cholesterol and may help lower high blood pressure.

Table 5.1

Vitamin A in Selected Foods*			
	1975	1997	Change
Broccoli	2,500 IU	1,543 IU	−38.3%
Cabbage	130 IU	133 IU	+2.3%
Cauliflower	60 IU	19 IU	−68.3%
Collards	6,500 IU	3,824 IU	−41.2%
Onions	40 IU	0	−100%
Parsley	8,500 IU	5,200 IU	−38.8%
Watercress	4,900 IU	4,700 IU	−4.1%

Vitamin C in Selected Foods*			
Broccoli	113 mg	93.2 mg	−17.5%
Cabbage	47 mg	32.2 mg	−31.9%
Cauliflower	78 mg	46.4 mg	−40.5%
Collards	92 mg	35.3 mg	−61.6%
Onions	10 mg	6.4 mg	−36%
Parsley	172 mg	133 mg	−22.7%
Watercress	79 mg	43 mg	−45.6%

Calcium in Selected Foods*			
Broccoli	103 mg	48 mg	−53.4%
Cabbage	49 mg	47 mg	−4.1%
Cauliflower	25 mg	22 mg	−12%
Collards	203 mg	145 mg	−28.6%
Onions	27 mg	20 mg	−25.9%
Parsley	203 mg	138 mg	−32%
Watercress	151 mg	120 mg	−20.5%

Iron Levels in Selected Foods*			
Broccoli	1.1 mg	0.88 mg	−20%
Cabbage	0.4 mg	0.59 mg	+47.5%
Cauliflower	1.1 mg	0.44 mg	−60%
Collards	1.0 mg	0.19 mg	−81%
Parsley	6.2 mg	6.20 mg	none
Onions	0.5 mg	0.22 mg	−56%
Watercress	1.7 mg	0.20 mg	−88.2%

Selected Nutrients in Broccoli*			
Calcium	103 mg	48 mg	−53.4%
Iron	1.1 mg	0.88 mg	−20%
Vitamin A	2500 IU	1542 IU	−38.3%
Vitamin C	113 mg	93.2 mg	−17.5%
Thiamin	0.10 mg	0.07 mg	−35%
Riboflavin	0.23 mg	0.12 mg	−47.8%
Niacin	0.09 mg	0.64 mg	−28.9%

* Based on 100 grams, edible portion
Source: USDA food composition tables

There has been much discussion about organic food, both throughout this book and in the media. Organic farming is certainly less harmful to the planet, but is it, as many claim, better for you? To begin, most people assume that "organic" means no pesticides. In January 1998 a *Consumer Reports* study using a variety of produce from around the country revealed that even organic fruits and vegetables are not pesticide-free. "But," the article stated, "the health implications of pesticides in food depend more on the amount and types of pesticides present than on the mere presence or absence of residues. And by this standard, the organic foods consistently had the least-toxic pesticide residues."[15]

Little research has been done on the nutritional value of local and regional foods, but organic produce has been the subject of many studies. One investigation, carried out in Chicago between 1995 and 1997 by trace minerals laboratory analyst Bob L. Smith, took a sampling of organic and nonorganic produce from local supermarkets and tested both types. An Internet source reported,

> The results are stunning, and should be a wake-up call to the whole world. The organically grown wheat had twice the calcium, four times more magnesium, five times more manganese and thirteen times more selenium as the non-organic varieties. The organically grown corn had twenty times more calcium and manganese, and two to five times more copper, magnesium, molybdenum, selenium and zinc. The organically grown potatoes had two or more times the boron, selenium, silicon, strontium and sulfur, and sixty percent more zinc. Overall, organically grown food exceeded conventionally grown crops significantly in twenty of the twenty-two beneficial trace elements.[16]

This is strong evidence that organic foods are actually better for us than conventionally grown produce, yet people tend to steer away from organic produce because it does not look as good as agribusiness produce. We want our food to be just like the food in the slick, colorful advertisements that surround us—perfect and blemish-free. Our food models are no different, in a way, from the supermodels who grace the pages of high-fashion magazines—an ideal presented to us by someone we do not know, in a way that makes us want to be or have what they are. Most of us do not look in the mirror and see Cindy Crawford or Mel Gibson looking back at us. Produce, however, is a different story. We can actually get produce that looks exactly like the fruits and vegetables we see in magazines, and studies show that we will buy it if we find it.

In *Food for the Future*, Patricia Allen cites a survey done in California:" Shoppers were shown photographs of perfect oranges and blemished oranges: 78 percent said that they would not buy the blemished fruit."[17] Just as too-thin models inspire anorexia and bulimia in those who seek to attain an unrealistic goal, perfect, blemish-free produce inspires a similar disorder in our agricultural system. As consumers we demand products of uniform color, size, and shape. Nutrition and pesticide use are generally not factors in our push for perfect food. Plants are bred specifically to exhibit the traits we seek in our food. Farmers will always opt for the food that sells best—either they can plant what will make them money or they can lose their farms. We end up with tomatoes that look great but taste like cardboard, peaches that look beautiful but are hard as rocks, and food that is grown in pesticide-drenched fields. Nature does not produce a perfectly round tomato or a bright red peach. On the contrary, nature is unpredictable, even chaotic—but what lies beneath that irregular exterior is almost always superior flavor and nutritional content. Interestingly, when the same people in the California supermarket study were told that the absence of blemishes on the oranges indicated the presence of pesticides, 63 percent of them decided they would rather buy the less-perfect-looking fruit.[18]

WE ARE WHAT WE FEED OUR CHILDREN

The examples we set for our children will determine their future health as well as the health of the planet. Unfortunately, thus far we are failing miserably. As *Eating Well* magazine reported in 1998, "When researchers from the National Institutes of Health surveyed over 3,300 children between the ages of two and 19 to determine their typical food intake, the results were alarming."[19] For starters, the diets of a full 16 percent of American children did not meet one single USDA recommendation. Only 1 percent met all criteria outlined on the USDA Food Pyramid. Childhood obesity, excess sugar consumption, and poor overall nutrition have been shown, as we have seen, to trigger disease. Attention deficit disorder (ADD) and attention deficit and hyperactivity disorder (ADHD) are two of the most commonly diagnosed illnesses among children. Doctors believe that diet plays at least a causal role in both ADD and ADHD. The current solution is Ritalin, an amphetamine-type drug—giving hyperactive kids a healthy dose of speed paradoxically slows them down. Some schools are so filled with pill-popping children that lines form outside the nurses' offices for the afternoon dosing. If what we are feeding our kids may be causing them to behave in ways that require them to take serious drugs, why not just change their diets? In 1996 the Feingold Association, a nonprofit group supporting a dietary approach to hyperactivity disorders, reported that "an increasing body of evidence supports diet as an effective treatment for ADD and ADHD." The report goes, on "Many of [these kids] can be helped by simply avoiding eating certain food chemicals. Changing brands, not necessarily the child's diet, can help them to avoid having to rely on powerful drugs. The research connecting diet and behavior often goes unread. For two decades, the scientific literature and parents have noted that certain synthetic, petroleum-based food additives act like drugs on certain children, impairing learning and triggering symptoms which look exactly like ADD and ADHD."[20]

Even if you are a parent to the 1 percent of children who met all of the USDA guidelines, and even if you have a healthy child, you may still have some problems. No matter how hard we try to feed our children a healthy, well-balanced diet, the foods we are feeding them can still make them sick. Governmental rules and regulations for acceptable levels of chemicals like pesticides, additives, and preservatives in our food supply are based on adult body mass. To add to the problem, children do not often eat the variety of foods adults eat, tending to focus on one or two types of fruits or vegetables. If they consume chemically adulterated foods, such as apples during the Alar scare, they are eating significantly more contaminated food per pound than the average adult would.

Because they are growing, children drink 2.5 times more water, eat 3 to 4 times more food, and breathe 2 times more air than do adults. In *Our Children's Toxic Legacy: How Science and Law Fail to Protect Us from Pesticides*, John Wargo reports, "Today, nearly 325 active pesticide ingredients are permitted for use on 675 different basic forms of food, and residues of these compounds are allowed by law to persist at the dinner table. Nearly one-third of these 'food-use' pesticides are suspected of playing some role in causing cancer in laboratory animals, another one-third may disrupt the human nervous system, and still others are suspected of interfering with the endocrine system."[21] Sandra Steingraber explains in *Living Downstream*, "In 1993, the National Research Council concluded that the current regulatory arrangement permits pesticide levels in food that are too high for children and infants. Childhood exposures to pesticides may lead to greater risks of cancer and immune dysfunction than exposures later in life, and one study found traces of sixteen pesticides in eight different baby foods purchased in U.S. grocery stores."[22] In 1992 Senator Edward Kennedy requested more information about levels of pesticide residues legally permitted to remain on foods purchased by consumers. EPA administrator William Reilly responded by admitting that levels for at least 60 of the permitted 325 food-use pesticides were above what is considered safe. If typical children's eating habits were taken into account, a significantly larger number of those same pesticides would be added to the list. Linda Fisher, an assistant administrator under Reilly, commented: "As you can see, infants and children are the two subgroups that typically receive the most exposure to pesticide residues in the diet as a percentage of body weight."[23] Chemical exposure for infants is of even greater concern—particularly since breast milk may be highly contaminated. Steingraber reports that "by 1976, roughly 25 percent of all U.S. breast milk was too contaminated to be bottled and sold as a food commodity."[24]

The key to eating right lies in education. If you understand what you are feeding your children, the chances of feeding them something harmful are greatly reduced. If you know that blemished organic produce is more healthful than its perfect-looking, flavorless cousins are, you can contribute to a healthier future. If children are taught how to eat properly, they will carry those lessons into adulthood and pass them along to their children. Some important educational programs are available to children, such as Alice Waters' Edible Schoolyard, CC2000s Adopt-a-School program, and the USDA's Team Nutrition; but, the best place a child can learn is at home, at the dinner table.

THE POWER OF MARKETING AND PUBLIC RELATIONS

With all the marketing dollars (nearly $4 billion in 1997) spent to influence our food choices, it can be difficult to see the truth clearly enough to understand the implications of our decisions. Plates (which incidentally are getting bigger to accommodate larger portion sizes) overflowing with food dance before our eyes on the television screen, and kids in magazines appear to be having the time of their lives while consuming unbelievable quantities of soda and sports drinks. Supermarkets and restaurants are answering the call, as Dan Goldberg comically and truthfully attests in "The Godzilla Effect: Way Bigger, Not Way Better."

> The other day I went into my favorite whorehouse, er, warehouse store and noticed that they have started selling fresh bagels. The bagels are sold out of bins, just like in the bagel store we went to when I was a kid, but the warehouse bagels were huge, at least three times the size of the ones I remember from my childhood. Indeed, when I saw someone come by with a forklift I assumed they had decided to buy a dozen.
>
> I turned a corner and spotted the warehouse's muffins and was reminded that these treats have also expanded to Chris-Farley-esque proportions, doubling in size and fat grams every five years over the past two decades. What was once a few bites is now a meal, or even a meal plus a snack.
>
> Hamburgers are getting bigger, too. The "Big Mac," which seemed huge when it first came out, looks anemic compared to the megaburgers everyone's purveying. And remember the height of down-home café overeating, the three-egg omelet? We recently went to such a café, and when I looked in the menu I nearly fell off my seat: they proudly serve five-egg omelets, plus your choice of two pork chops or nine sausage links or two-eight-ounce steaks. Don't worry, hash browns and toast or biscuits are also included....
>
> I pulled up at a light the other night and saw a guy in a beat-up pickup next to me eating Kentucky Fried Chicken. His face was the size and color of an Easter ham (the "serves 30" variety) and just as greasy. I watched him strip the meat off a drumstick in one bite, then take a big pull from a soft drink the size of a small garbage can. And I thought: If there were truth in advertising, this guy is who you'd see in just about every food commercial. Even the ones for bagels.[25]

As a matter of fact, Goldberg's article, written in 1998, is already out of date. ABC recently broadcast an episode of *20/20* that reported that omelets have grown to include eight eggs, and that McDonald's has answered the "bigger is better" call with an even more monstrous version of the Big Mac. Amazingly, at the very same time that we are being encouraged to eat these Alice-in-Wonderland-sized portions, fat-free—but not calorie-free, a fact that the marketing gurus are paid highly to keep from us—products are filling up the shelves with lightning speed.

Just as quickly, unanswered questions about our health fly in and out of our consciousness. Is it okay to have an extra serving of beef at dinner? Should we allow our children to drink soda? If fat is bad, is it better to eat no fat at all? From prehistoric times until literacy became widespread, most food knowledge was passed down from one generation to the next by word of mouth or by example. Mothers taught their children which foods should or should not be eaten together, the ways in which things should be cooked, and which foods were dangerous or unhealthy. Lessons in food history and safety were learned as part of the culture and ritual of feeding. This culture that surrounds food has, for the most part, been lost to a world of fast food and frenzied frozen dinner preparation.

A 1998 home meal replacement study done in Chicago reports that "the average family rarely sits down to eat together, and relies heavily on home meal replacements to get them through their busy week." The study went on to say that "97 percent of the respondents reported using alternatives to home-cooked meals at least one night during the week."[26] The food service industry's vision of the future is one that sees people cooking with fewer raw and more processed products. Fewer meals will be eaten at home, and those that are, if predictions hold true, will have been purchased pre-made, from food service establishments. Sitting down to "break bread" together will, and to some extent already has, become an insignificant part of another activity, such as a meeting, social, or sporting event. Meals are becoming more solitary in nature—eaten alone in an office, on the street, in a car, or in front of the television. Irena Chalmers addressed this issue in a 1995 speech in the Dominican Republic at the Caribbean Conference on Sustainable Tourism, cosponsored by the United Nations.

> How rapidly food habits have changed. At the turn of the century the average woman had five children and spent five hours a day in the kitchen. Now the average family has 1.8 children and spends barely 15 minutes to prepare dinner. If a microwaved packaged dinner needs to be stirred or turned more than once, it is not acceptable to the market. Even this minimal involvement is considered to be too much trouble. We are fast becoming a nation that is cooking illiterate. Today's Moms, who have been working away from home all day, often feel guilty about having "abandoned" their family, even though their salary is essential to maintaining the household. The decision about what to eat and where to eat is no longer the decision of the mother, but the children who choose the fast food chain, not on the quality of the food—or lack of it—but on the basis of the toys that are given away. Last year advertisers spent $500 million for television ads directed at children.[27]

One of the challenges we face in trying to establish good, safe, healthy, nutritious eating patterns for our children is to overcome the effect of advertising dollars being spent not only on TV and in print, but also in the classroom. Millions of dollars are being spent by corporate food service behemoths to entice children at a very young age—their goal, to shape future eating patterns with no regard for children's health. In *When Corporations Rule the World*, David Korten describes this problem:

> Corporations are now moving aggressively to colonize the second major institution of cultural reproduction, the schools. According to Consumers Union, 20 million U.S. schoolchildren used some form of corporate-sponsored teaching materials in their classrooms in 1990. Some of these are straightforward promotions of junk food, clothing, and personal-care items. For example, the National Potato Board joined forces with Lifetime Learning Systems to present "Count Your Chips," a math-oriented program celebrating the potato chip for National Potato Lovers' Month. NutraSweet, a sugar substitute, sponsored a "total health" program. Corporations have also been aggressive in getting their junk foods into school vending machines and school lunch programs. Trade shows and journals aimed at school food-service workers are full of appeals such as: "Bring Taco Bell products to your school!" "Pizza Hut makes school lunch fun." Coca-Cola launched a lobbying attack on proposed legislation to ban the sale of soft drinks and other items of "minimal nutritional value" in public schools. Randal W. Donaldson, a spokesman for Coca-Cola in Atlanta, said: "Our strategy is ubiquity. We want to put soft drinks within arm's reach of desire. We strive to make soft drinks widely available, and schools are one channel we want to make them available in."[28]

Any parent of young children knows the pull of slick advertising. Corporations have a serious incentive to work against a parent's struggle to feed their children healthy, nutritious food.

For most of us, the media is our direct link to the knowledge we acquire about the health and safety of our food supplies. In January 1998 *Food Insight* newsletter published a story titled "In the News: A Snapshot of Food and Nutrition Reporting." The article, which was based on a study commissioned by the International Food Information Council (IFIC) Foundation, states that "the media often neglected to provide the context necessary for consumers to make informed choices about their own food selections. For example, news reports with advice on what foods to eat more of, rarely specified how much, how often, or to whom the advice applied."[29] The article goes on to say that culinary groups and food producers are quoted as experts more often than scientists and government officials. Food-borne illness was the number one newsworthy culinary topic in 1997. Coverage of fat consumption dropped by 50 percent from two years earlier. The *Food*

Insight report revealed that nearly 30 percent of news stories on functional foods neglected to mention important details, such as the amount, frequency, and cumulative or noncumulative effects of consumption. They also fail to address differences found among various groups, such as pregnant women, children, and the elderly.

It does not help that even the experts are unable to agree. In 1998 *Newsweek* reported that Dr. Benjamin Spock "advocated raising kids on a vegan diet in the new version of his 50-million-copy best seller *Baby and Child Care*. The seventh edition, published weeks after his death in March, reads, 'We now know that there are harmful effects of a meaty diet' and 'I no longer recommend dairy products after the age of two years.' But pediatrics expert T. Berry Brazelton was aghast at news of his friend's words, telling *The New York Times* the guidelines were 'absolutely insane' and that kids need the protein and calcium of milk and meat."[30]

PROCESSED FOODS: WHAT'S IN WHAT WE EAT

Our processed foods are packed full of preservatives, food coloring, fake fat, fake sugar, fake salt, and a whole host of other chemicals—whether we want them or not. Today these things are so common in the foods we eat that most of us do not even think to question them. Sometimes, even when we try to make informed decisions about what we are eating, the label can be misleading. The product label is a vital link between the consumer and his food. Without proper and accurate labeling it is impossible to know what to feed our friends and families.

As a chef I pride myself on knowing about food. For twenty-six years it has been my life, yet it is only during the past five years that I have truly come to question our food supplies. Joining CC2000 and going to my first conference in Puerto Rico in 1996 forced me to take a second look at everything I was doing in the kitchen and everything I was buying. The more I learned, the more concerned I became and the more complicated the issues seemed. The way food additives appear to sneak into our food supply is astonishing. Olestra is a case in point. I first heard about it from Joan Gussow at a CC2000 retreat and watched in horror as its popularity gained momentum. In 1998 *Newsweek* reported on the growing usage and ramifications of this fat substitute:

> Olestra, or sucrose polyester, is a fat assembled from natural sources that has been chemically processed to make it indigestible. Because Olestra passes right out of the body, foods made with Olestra derive none of their calories from fat. But Olestra is fat, if a weird form of it. As it moves through the gut, it attracts fat-soluble nutrients and carries them out of the body. Those nutrients include vitamins A, D, E and K and some carotenoids, which are substances found in fruits and vegetables that help protect against heart dis-

ease and many cancers. Walter Willett of the Harvard School of Public Health
estimates that if consumption of Olestra snacks becomes widespread, Americans may experience up to 50,000 more cases of cancer and heart disease
every year.[31]

We have known from the start that Olestra can have unsavory minor
side effects, and many experts believe that insufficient research has been
done to understand the long-term effects of the product. Yet sales of snack
foods made with this product are booming. In 1998 the *Los Angeles Times*
reported that "buoyed by a second positive regulatory review, salty snacks
made with the Olestra fat substitute are turning into one of the decade's
most successful new food-product introductions. A survey conducted by
Chicago-based Information Resources Inc. for Frito-Lay Inc. was cited as
showing that the Dallas-based snack-food company sold $58 million of its
WOW! Brand chips made with Olestra during its first two months in the
market. Lynn Markley, spokeswoman for Frito-Lay, was quoted as saying,
'There is no question it has been the most successful product launch of the
1990s.'"[32] Just tell consumers that there is no fat in something, and they
flock to the supermarket like lemmings. If it tastes good, it is welcomed with
open mouths.

Fat substitutes are not the only substitute additives to find their way into
our food supply. Sugar substitutes have been in the marketplace for many
years and continue to grow in popularity. Leading the pack is Monsanto's
NutraSweet, but it faces new competition from Johnson & Johnson's latest
artificial sweetener, sucralose. Approved in 1998, sucralose is 600 times
sweeter than sugar. The Associated Press reported in April 1998, "Johnson
& Johnson's sucralose is the only artificial sweetener made from sugar. Scientists changed sugar's molecules to make the sweetness more intense but
not allow it to be absorbed. Unlike other sweeteners, it passes straight
through the body."[33]

Strangely, sucralose has some of the same attributes as Olestra and perhaps some of the potential problems, but as the product has only been on
the market for less than a year, it may be quite some time before the problems come to the public's attention. In May 1998 Veryfine Products, Inc.,
makers of fruit juices sold throughout the country, introduced a "diet" juice
using the above-mentioned sucralose. Where does it stop? Will we soon
have food that is entirely made of fake ingredients that pass, like stealth
technology, through the digestive tract?

LABELING

Anyone who has ever tried to read a label knows that it can be a frustrating experience. In an effort to simplify things, the FDA has standardized
the food label to highlight nutritional information, including amount per

serving of saturated fat, dietary fiber, and "other nutrients of major health concern," such as sodium, caffeine, protein, and vitamin content. Required elements are total calories, calories from fat, total fat, saturated fat, cholesterol, sodium, total carbohydrate, dietary fiber, sugars, protein, vitamin A, vitamin C, calcium, and iron. Voluntary items include calories from saturated fat, polyunsaturated fat, monounsaturated fat, potassium, soluble fiber, insoluble fiber, sugar alcohol, other carbohydrate, percent of vitamin A present as beta-carotene, and other essential vitamins and minerals. If a specific product claim is made about the listed voluntary items, labeling in that case becomes mandatory. This kind of labeling is required for most food products, but is voluntary on the twenty most frequently consumed raw fruits, vegetables, and fish, as well as the forty-five best-selling cuts of meat. Confused yet? On top of that, the descriptive terminology can be deceiving.

Nutrition Facts

Serving Size ½ cup (114g)
Servings Per Container 4

Amount Per Serving

Calories 90   Calories from Fat 30

	% Daily Value*
Total Fat 3g	5%
Saturated Fat 0g	0%
Cholesterol 0mg	0%
Sodium 300mg	13%
Total Carbohydrate 13g	4%
Dietary Fiber 3g	12%
Sugars 3g	
Protein 3g	

Vitamin A 80%	•	Vitamin C 60%
Calcium 4%	•	Iron 4%

* Percent Daily Values are based on a 2,000 calorie diet. Your daily values may be higher or lower depending on your calorie needs:

	Calories:	2,000	2,500
Total Fat	Less than	65g	80g
Sat Fat	Less than	20g	25g
Cholesterol	Less than	300mg	300mg
Sodium	Less than	2,400mg	2,400mg
Total Carbohydrate		300g	375g
Dietary Fiber		25g	30g

Calories per gram:
Fat 9 • Carbohydrate 4 • Protein 4

For instance, the word *free* on a label, as in "fat free," actually means that the product contains less than half a gram of fat, not no fat. "Low fat" means no more than 3 grams of fat per serving. It does not necessarily mean the product contains fewer calories. For example, a typical candy bar might have 200 calories. A low-fat alternative may still contain 150 calories, which is probably considerably more than you might expect. You may also be surprised to learn that low-fat products often contain more sugar than their full-fat counterparts. Your fat intake has been lowered, but you may actually be consuming more calories. Nonfat frozen yogurt may seem like a healthy alternative, but it often contains the same number of calories as regular frozen yogurt and the potential for the same weight gain. It is easy to become confused when shopping for healthful "natural," or "whole" foods. "Natural" foods, though their designation implies that they are completely unadulterated, can actually be minimally processed. "Whole" foods may be in their original form or could be minimally processed and might be organic, certified organic, transitional organic, or organically grown. Until the federal government produces its regulations, we might all have to shop with spreadsheets and decoder rings to figure out what we are buying.

But labeling is not the only area of confusion. As a chef I am often called upon to produce vegetarian, vegan, no-fat, or other entrees for people who have dietary restrictions or personal dietary preferences. However, during the research for this book I came across information that made me conscious of how little most of us understand our food supplies. One of the most surprising things I have learned is that vegetables, including organic vegetables, can be grown with fertilizer made of cattle by-products. In fact, bone and blood of slaughtered animals is used extensively in vegetable production. This could prove disturbing to vegans and vegetarians who eat organic vegetables precisely to avoid eating meat and meat by-products, and it is a chilling reminder of how pervasive cattle farming is in our diets—vegetarian or not, organic or not. Michael Ableman comments in *On Good Land: The Autobiography of an Urban Farm*, "There is a strange paradox in this for those who think that by eating organic vegetables they are absolved of involvement with these deplorable places [cattle farms]. Many organic farmers rely heavily on blood meal, bonemeal, and purchased composts to augment their soil fertility program. These fertilizers are the by-products of the worst of the livestock industry. The manures originate from feedlots where animals are jammed together in squalor, the blood and bonemeal from mass production slaughterhouses. I am not blameless in this; I sometimes use blood meal on crops that consumers expect to be large in size."[34]

Even a term as simple as "fresh" is ambiguous when it comes to food. As you already know, chicken can be labeled fresh, even when held at 26°F—virtually frozen solid. Some fruit and produce, like apples, potatoes, and squash, may be labeled fresh even if washed in an acid solution and held for up to a year. The "fresh" fish you purchase may be several weeks old.

In fact, when fishing boats go to sea for ten days or more, harvested fish is held in heavily salted water at temperatures close to freezing for many days prior to being sold. And it is still considered fresh when it hits your table up to three weeks later. Typically, meat is aged before it is sent to your supermarket, but it can still be considered fresh when more than six weeks have passed after slaughter. Aging meat intensifies the flavor, so this is not necessarily bad, but I think this is a good example of how we perceive words like *fresh*, which to most of us means just picked, harvested, or slaughtered.

Improperly handled or labeled animal products may simply go bad before you get a chance to cook them. They could also be dangerous if not properly cooked. I wondered if fruits and vegetables also tend to lose nutritional value over time, so I asked Laura Pensiero, author of the *Strang Cookbook for Cancer Prevention*, if she had any insight. She told me that although research on specific vitamins has been done, there is still no definitive answer as to whether the nutritional healthfulness of fruits and vegetables decreases over time. However, when she posed a similar question to "Abby Block, who wrote the position paper for the ADA on functional foods and phytochemicals. . . . She said that her advice to people . . . in an urban setting, like New York City, unless you are shopping at a farmer's market, your best bet would be to buy frozen vegetables. Because if you go to the Food Emporium, or to a supermarket you can almost guarantee that you are getting ten-day-old vegetables with less nutrients in them." Eliot Coleman also told me that he was frustrated with the lack of available research. He even went to the USDA and the National Fresh Fruit and Vegetable Association, and not only do they have no information on nutritional changes in older produce, but to date neither one has even given him a definition of *fresh*.

PHARMACEUTICAL AND NUTRACEUTICAL DIETARY SUPPLEMENTS

Dietary supplements are not just the vitamins you take with your breakfast in the morning. They have long been a part of our food supply (in enriched bread, for instance), but now they are taking on an entirely new dimension, and the food industry is welcoming dietary supplements as a new way of making money. In an article published in early 1999, *Food Product Design* reported that the food service industry needs to gird itself for battle as it enters the nutraceutical/functional-foods fray. The potential prize could be enormous—estimates for the category are in the billions. The article went on to explain that labeling issues will be of foremost importance in terms of marketing future products of a "nutraceutical" nature.[35]

Manufacturers are being permitted by the U.S. government to decide whether to market their products as foods or as dietary supplements. This is particularly significant because the Nutritional Labeling and Education Act has set forth stringent restrictions regarding health claims made by anyone marketing food products. The legislation governing dietary supplements (Dietary Supplement Health and Education Act, or DSHEA) allows manufacturers to make claims that are "truthful and nonmisleading statements of effect on a body system, organ, or function, provided that the context in which the claim appears does not suggest that the product treats, prevents, or mitigates a disease." The DSHEA includes products that come in pill or liquid supplement form as well as drinks and health bars that fall more under the general heading of "food."

The result of this legislation can be confusing. It seems that a carbohydrate-based bar largely consisting of food ingredients (and often consumed by busy people to replace lunch) is a supplement. However, a fat-based spread designed as a carrier for cholesterol-reducing sterols falls under the category of food. *Time* health columnist Christine Gorman, reporting on this quandary in May 1999, had this to say:

> I don't ask a lot from food. It should taste good and be reasonably good for me. But more and more these days we're encouraged to view the grocery store as a medicine chest. There are tofu and yams for hot flashes. Ginseng tea for energy. Stewed tomatoes to prevent prostate cancer. So when I heard about Benecol® and Take Control®, the new margarines that are supposed to lower cholesterol levels in the blood, I didn't exactly smack my lips in anticipation. Still, I figured, given how much heart disease there is in the U.S., they deserved a look. Both spreads come with pretty good scientific credentials . . . randomized, controlled trials show that folks with mildly elevated cholesterol levels (between 200 mg/dl and 240 mg/dl) who ate roughly two tablespoons of Benecol a day decreased their level of ldl, the "bad cholesterol," about 14 percent. . . . Neither product is calorie free: both contain mostly polyunsaturated and monounsaturated fats. So if you eat too much, you'll be sure to gain weight, which can raise your cholesterol levels all by itself. Nor will the spreads do you any good if you spend all your waking hours in front of a television or computer screen. The biggest catch is that you have to eat three servings a day for the rest of your life. When test subjects stopped using the spreads, their cholesterol levels crept back up within a week. It's sort of like taking medicine—medicine that costs as much as $5 for a week's supply. Let's face it: even though these designer margarines appear to be safe now, who knows what we'll learn after hundreds of millions of people have eaten them? If you want to be part of a giant experiment that could very well save you from a heart attack but might expose you to unknown risks, be my guest.[36]

In addition to food supplements with questionable regulation, we are also experiencing increased consumption of herbal supplements, which are regulated very differently. Herbal products do not have to be proven safe or effective, and there is no guarantee that the herbal product is what its label advertises it to be. While prescription and over-the-counter drugs and food additives must meet the FDA safety and effectiveness requirements, herbs marketed with medical claims bypass these regulations. This may lead to misleading claims, unhealthy supplements, and too many people believing that a quick fix by supplement can replace good eating habits as a path to health. For the time being, this industry, which is going to grow exponentially in the next few years, will have to be policed by the consumer. This is truly a case of "buyer beware."

"WHAT A MAN SOWS, THAT SHALL HE AND HIS RELATIONS REAP"

When British poet Clarissa Graves (1892–1985?) wrote the above line, she was writing about family relationships, not farming. She could just as well have been discussing our relationship to the planet. Indeed, the idea of reaping what one sows is age-old farming wisdom. However, somehow, somewhere along the way, many of today's farmers have been making a full-blown effort to unlearn the meaning of this principle, and this is having a devastating effect on the land, animals, and the population.

Antibiotic Resistance

Animal husbandry is a case in point. Everything from cattle to fish is being raised, slaughtered, and brought to market in ways that are increasingly dangerous to consumers. To begin with, we are using antibiotics not to fight disease but to increase growth potential in various animals, including cows, fish, and chicken. As a result, many antibiotics no longer fight disease as they once did. In fact, over the past few years, the strain of E. coli O157:H7 has formed resistance to all but one antibiotic. One reason is that cattle are routinely dosed antibiotics to promote rapid weight gain. The faster an animal grows, the faster it goes to slaughter, and the more money a rancher will make. Amazingly, of the 50 million pounds of antibiotics produced every year in the United States, about 40 percent is given to animals. Use of these medicines can foster antibiotic resistance in bacteria common in animals, such as salmonella and E. coli, which cause disease in humans. As biologist Rebecca Goldberg has noted, "Antibiotics are probably the most controversial chemicals used in aquaculture. As of 1991, roughly 50 antibacterial drugs were being used in aquaculture worldwide." She goes on to say that the "use of antibiotics in aquaculture may contribute to extremely troubling and increasingly common problems with bacteria that cannot be controlled with commonly available antibiotics."[37]

The American Society of Microbiology has singled out the use of antibiotics in aquaculture as "potentially one of the most important factors leading to the evolution of antibiotic-resistant bacteria."[38] Resistant bacteria can be passed on to people through the food supply. Infections of antibiotic-resistant bacteria can be extremely difficult for physicians to treat, and may result in extended hospital stays or even death. The Centers for Disease Control (CDC) report that salmonella poisoning has increased sixfold in the past twenty-five years, and that forty times more strains of tetracycline-resistant salmonella now exist. In the May 1999 *Time*, Dick Thompson reported that the *New England Journal of Medicine* connected antibiotics used to treat poultry to disease resistance in humans:

> More than 19 million lbs. of antibiotics are fed to cattle, pigs and chickens each year as they amble toward the dinner table. At the same time, doctors treating meat-eating humans have seen a steady and alarming increase in infections resistant to these same antibiotics. Is there a link? Scientists and consumer activists long suspected that there was but were never able to prove it. Now they can. In the first study to connect antibiotic resistance in humans directly with the food we eat, a group of Minnesota public health specialists reported in last week's *New England Journal of Medicine* that an eightfold increase in drug-resistant food poisoning among Minnesotans directly followed the approval and use of the same drug in chickens.[39]

The FDA has become so concerned about this situation that in 1999 they revised their guidelines in the hopes of diminishing the routine use of these drugs. In March 1999 the *New York Times* explained the FDA's new guidelines:

> The goal of the revision is to minimize the emergence of bacterial strains that are resistant to antibiotics. . . . Of particular concern to scientists is that recent studies have found bacteria in chickens that are resistant to fluoroquinolones, the most recently approved class of antibiotics and one that scientists had been hoping would remain effective for a long time. A crucial component of the new guidelines will be the requirement that manufacturers test certain new livestock drugs for a tendency to foster the growth of resistant bacteria that could prove harmful to people. . . . Last May, a team from the Centers for Disease Control and Prevention reported in the *New England Journal of Medicine* that the prevalence of a salmonella strain resistant to five different antibiotics increased from 0.6 percent of all specimens from around the country tested by the centers in 1980 to 34 percent in 1996. Similarly, drug resistance in campylobacter bacteria rose from zero in 1991 to 13 percent in 1997 and 14 percent in 1998, said Dr. Fred Angulo, a medical epidemiologist in the foodborne and diarrheal disease branch at the Centers for Disease Control and Prevention. He said epidemiologists were "alarmed" by the campylobacter figures, because the resistance is to fluoroquinolones, the very class of drug that the F.D.A. is trying hardest to preserve.[40]

Giving so-called therapeutic doses of antibiotics to animals that we consume is a bad idea. Treating a sick animal is reasonable; dosing animals with antibiotics for weight gain is unreasonable, especially if it puts us, our families, and our future at risk.

Mad Cows

Mad cow disease hit Britain like a volcanic eruption. Media reports in the 1990s indicated that ten people were dead from eating beef. In 1996, the British Ministry of Health confirmed the cause to be "bovine spongiform encephalopathy (BSE)," a degenerative brain disorder of cattle. Almost immediately sales of British beef to other parts of the world were halted as the country's own sales fell off sharply. The disease has grown out of a tradition in animal farming that allows the carcasses of sick beef, lamb, and pork animals to be ground into animal feed for both beef and dairy cattle. It seems crazy that we would feed animal products to vegetarian animals—even crazier that we would feed beef animals back to beef animals. Shortly after news of the outbreak surfaced, I was working as consulting chef for a restaurant in London and was surprised to learn that the owner was demanding American beef for his guests. There was a general feeling among the British at that time that mad cow disease would never occur in people eating American beef because laws regulating animal production were much more stringent in the United States. As far as the owner of the restaurant was concerned, nothing else would do, but I found myself wondering whether their trust in American regulatory agencies was justified.

In 1998 *Atlantic Monthly* reported that mad cow disease could certainly happen here:

> The chain of reasoning that should make us worry begins with the fact that the economics of our modern meat and milk industries dictate that many farm animals get food supplements derived in part from rendered animal protein. The rendered animal protein they eat may expose them to the infectious agent, which is thought to have the potential to cross the species barrier between animals, even into human beings. The disease is 100 percent fatal, and in human beings takes up to thirty years to manifest symptoms. The FDA banned the practice of feeding to cattle or sheep almost anything containing the meat, bone, or fat of mammals other than pigs and horses. Some observers are skeptical of this ruling, in part because the FDA has allotted the equivalent of only seventeen full-time inspectors to 14,000 facilities involved in feed production and rendering—the process by which animal by products, such as meat, bone, and fat, are melted down and separated. Another objection is that the feeding of cows' blood to cattle continues to be permitted. The skeptics argue, too, that there is no way to prove that the species barrier cannot be surmounted, and the current rules allow even animals known to be infected to be fed to pigs and chickens, which can in turn be fed to us. Or they may be fed to cattle and sheep.[41]

Whatever the transmission path, feeding dead diseased animals to live animals seems a bad practice.

In April 1999, NBC news correspondent Jim Avila reported that in June a warning from the government could forever change the way millions of Americans eat lunch:

> The scheduled warning singles out packaged, ready-to-eat meats from hot dogs to sliced cold cuts for being vulnerable to infection by a rare, but often lethal germ called Listeria. The problem is so perplexing and potentially harmful that the USDA is now writing new guidelines for both the makers of ready-to-eat meat and those who buy it. Those recommendations include warnings that children, older Americans, people with immune and medical problems and especially pregnant mothers not eat hot dogs or lunchmeats unless they are recooked at home. Specific warning labels advise pregnant mothers of the risk to fetuses. And for the manufacturers, the government will recommend repasteurization (cooking the meats twice), including once in their packages. The problem is that inside a meat plant, Listeria strikes after the food is cooked and before it's packaged. This year alone, there have been 10 major meat recalls.[42]

Americans consumed nearly 75 pounds of beef per person in 1998. I would not suggest that you should stop eating beef altogether, particularly if you enjoy it, but it would certainly be wise to consider the source of your meat. Certain producers raise healthier beef than others do. Coleman Natural Beef and Laura's Lean Beef are two companies raising beef without antibiotics, hormones, and adulterated feed. Coleman also allows his animals to graze freely on open rangeland. A study done in 1999 even suggested that pasturing cattle for two to three weeks prior to slaughter as opposed to fattening them in feedlots may prevent a high percentage of contaminated beef. In early 1999 the USDA announced that it would allow all of the national organic certification bodies to certify beef and that antibiotics and hormones would not be allowed in beef labeled organic. Hopefully this will bring healthier, safer beef into the marketplace. The government's ruling allowing irradiation for meat products (with no identifying labeling) in concert with agribusiness's technology, on the other hand, is trying to cure the problem the easy way.

Hormonal Hijinks

It has recently come to my attention that young women are reaching puberty earlier, older women are entering menopause sooner, and fewer male babies are being born than ever before. At first glance these three facts may seem somewhat unrelated, but many researchers believe that all are partially the result of a combination of the hormones administered to animals we eat and the pesticides and pollution that have become a part of our food supplies. In 1997 a study published by the Pediatric Research in Office Settings Network of Pediatrics reported that "young girls are maturing earlier than in the past, and different races reach puberty at different ages. In a recent study of 17,077 girls (9.6 percent African-American; 90.6 percent white) seen by 255 clinicians in 65 pediatric offices in the United States, researchers found that girls are developing breasts, growing pubic and other body hair, and menstruating earlier than the standards used by most clinicians. Although there are no clear-cut reasons behind these phenomena, possible explanations . . . are increased use and exposure to estrogenic chemicals found in hair-care products, plastics and insecticides."[43] The *Green Guide*, published by Mothers & Others for a Livable Planet, similarly refers to studies done at the University of North Carolina and reported in the April 1997 edition of *Pediatrics*, "We have growth hormones in meat and milk, plastics containing estrogen-like substances in dental sealant used on kids' teeth, and plastic wrap that may have an estrogenic effect used to wrap sandwiches for kids' lunches. PCBs and DDE (a breakdown product of the pesticide DDT) may indeed be associated with early sexual development in girls."[44]

Historically, 125 males have been conceived for every 100 females, and about 105 boys are born for every 100 girls. Over the past few decades, those numbers have changed, according to an article in *Scientific American*: "In the U.S. in 1970, 51.3 percent of all newborns were boys; by 1990, this figure had slipped to 51.2 percent . . . in Canada the decline has been more than twice as great and similar long-term drops have been reported in the Netherlands and Scandinavia. Some researchers believe pollution may be the culprit. A recent article in the *Journal of the American Medical Association* notes that high exposures to certain pesticides may disrupt a father's ability to produce sperm cells with Y-chromosomes—the gametes that beget boys. Other toxins may interfere with prenatal development, causing a disproportionate number of miscarriages among the frailer male embryos."[45]

Environmental activists and some physicians have expressed concern about this issue for many years. Rachel Carson first brought it to public attention in *Silent Spring*, in which she reports that "research from Scotland, Michigan, Germany, and elsewhere indicates that [pesticides] lead to reduced fertility, testicular and breast cancer, and malformation of the genital organs. In the United States alone, as the tide of estrogen pesticides has crested in the past twenty years, the incidence of testicular cancer has risen by approximately 50 percent. The evidence also suggests that, for reasons not yet understood, there has recently been a worldwide drop in sperm counts of 50 percent."[46] Today the American Medical Association believes this drop may be related to agricultural pesticide use. Carson's book warned of these hazards over a quarter century ago, yet at the dawn of the millennium we still are not taking the threat seriously.

FOOD SAFETY

Whether it is contaminated at the farm, at the processing plant, or in your own home, tainted food is the worst kind of betrayal. Cases of food-borne illness are at an all-time high. Many experts believe that it is not only possible but also quite probable that during the early part of this century we will witness an outbreak of food-borne illness that will sicken hundreds of thousands of Americans, with an unprecedented death toll. Some of the most recent outbreaks may foreshadow things to come.

In the nation's largest beef recall as a result of E. coli contamination, 25 million pounds of meat were recalled due to a 1997 outbreak that sickened fifteen people in Colorado and one in Kentucky. Hudson Foods and two of its employees were charged with lying to regulators about the quantity of meat that may have been tainted during that recall, the Associated Press reported; the company lost its account with Burger King and eventually changed its name. Another outbreak, this time at a Chicago meat-processing firm, prompted the voluntary recall of three weeks' production of frozen ground beef patties thought to be contaminated with E. coli bacteria.

In 1998 the Associated Press reported that Bil Mar, a subsidiary of Sara Lee, was responsible for a bacterial outbreak that killed four people and made thirty-eight others sick. The bacterium listeria was found in hot dogs, but it was impossible to know whether it originated at the Bil Mar plant or was brought in from outside sources. As a result of the outbreak 15,000 pounds of turkey franks, distributed in thirteen states, were recalled. A class action lawsuit was brought against Sara Lee that same year, but the outcome is still pending. Yet another 1998 Associated Press story reported that the Michigan Department of Agriculture discovered that Herrud Franks made with chicken, pork, and beef were potentially contaminated with listeria.

Both eggs and egg-based products such as ice cream have been making people ill in the past few years, and in 1998 *Reuters* reported that almost 900,000 Americans are sickened each year by eggs contaminated with salmonella. Doctors however, report much smaller numbers. In 1994 the Centers for Disease Control (CDC) announced that Schwan's ice cream was behind a food-borne illness outbreak that they estimate sickened 224,000 people. Trucks used to transport Schwan's premix, which is also prepasteurized to avoid salmonella poisonings, had previously been used by another company to transport unpasteurized eggs, and were improperly cleaned.

Although adulterated eggs, beef, and chicken are the foods we most commonly hear about in the news, we are increasingly becoming ill from foods that we might never consider to be hazardous. In 1998 the *New York Times* reported that in 1996 sixty-one people had become ill and twenty-one were hospitalized, one critically ill. All they ate was lettuce. A label claiming the lettuce was fresh and "pre-washed" made consumers feel as though they were getting clean, safe food. In this case, however, the open tub in which the lettuce was washed was downwind from a pasture where cattle were grazing. Wind carried dried bits of manure into the lettuce-washing facility, and with it came enough *E. coli* to make all those people ill. The problem was exacerbated by lack of sanitation on the part of farmworkers, who tracked more manure into the facility and did not engage in proper hand-washing practices. Another *E. coli* outbreak from an unexpected source occurred in 1996, when 66 people fell ill after drinking Odwalla apple juice. Odwalla juice had been thought of by consumers as a good, healthy juice. Mothers felt comfortable letting their children drink Odwalla, and juice manufacturers were doing everything they could to produce a high-quality product. The true agony of food-borne illness is always most evident in children. One little girl, Anna, quickly became critically ill; in two weeks her kidneys stopped functioning and her brain became clogged with dead blood cells. Eventually her heart failed, and she died. The *New York Times* reported in 1998 that "authorities say that while Odwalla was marketing its products as premium sources of nourishment, the fast-growing company had significant flaws in its safety practices. . . . Interviews with former Odwalla managers and company documents show that in the weeks before the outbreak, Odwalla began relaxing its standards for accepting blemished fruit."[47] While no one is certain how *E. coli* ended up in the Odwalla apple juice, some have theorized that Odwalla's growers may have sent the company apples that had fallen to the ground. Since the company had changed its requirements about not accepting blemished fruits, there was no way of knowing whether the apples had all been picked from the tree (as per Odwalla's specifications), or if they had fallen to the ground and begun rotting there. It is believed that when the apples hit the ground they were very likely contaminated through contact with deer or other animal feces. After the outbreak, Odwalla began pasteurizing its juices and smoothies.

In yet another unlikely scenario, in 1998 salmonella bacteria was discovered in boxes of toasted oat cereal manufactured by Malt-O-Meal. After the CDC determined that the cereal was the cause of 188 cases of food poisoning in eleven states, the company voluntarily recalled nearly 3 million pounds of cereal. The source of the bacteria is thought to have been an oven that had been broken for a number of months, not warming to a high enough temperature to kill bacteria. A vegetable spray used to keep the toasted cereal from sticking in the oven had built up over time and gone rancid, later contaminating the cereal.

Not all food-borne illness originates in restaurants or industrial food service operations. In fact, some of it can be found right in our own homes. In speaking with friends and employees, I have often found that people's notions of safe food handling may be seriously in error. Most of us probably have memories of a childhood Thanksgiving when due to lack of refrigerator space the leftover turkey trimmings were left in a cool place overnight. It makes me wonder how many of our stomachaches as children were really illnesses caused by improperly handled food. A 1998 article in the *Orlando Sentinel* reported that "Audits International, a food consultant, reviewed 106 household kitchens in 81 cities. Its findings: Only one kitchen passed the review. The others had problems with hand washing, storing food at the right temperatures and tainting ready-to-eat food with dirty utensils. 'Our study says 99 percent of the households we looked at would not be acceptable if they were operating as a restaurant.' "[48]

Food Safety Tips

A great deal of food-borne illness can be prevented through safe food-handling techniques. Pathogens only need three living conditions in order to survive and multiply: water, protein, and a moderate pH. Food that is high in protein and has a significant water content and a moderate pH makes a much better host for pathogens than food with little protein and an acidic pH. Temperatures between 40°F and 140°F are commonly referred to as the "danger zone," because most bacteria multiply most rapidly in that range. Heating to temperatures above 140°F will destroy the majority of pathogens, while freezing or cooling quickly to temperatures below 40°F will primarily inhibit bacteria growth until such time as a product is cooked.

Storing Foods

Restaurateurs are accustomed to keeping an eye on the internal temperatures of their freezer and refrigeration units, but home cooks rarely take the time to check how well their refrigerators are working. Placing an appliance thermometer just inside the refrigerator door will help you to keep track. Generally speaking, refrigerators should be maintained at 36 to 40°F. Note that different parts of your refrigerator are cooler than others, and for that reason, it is best not to store perishables, like eggs, on the refrigerator door, which normally has the highest holding temperature. Freezers should be kept at 10°F or below.

Cooling Foods

Most people make a meal, serve it, and then let the leftovers sit out until they reach room temperature. This is a very dangerous habit, especially if you are accustomed to letting food sit out for more than a few hours or overnight. The less time foods spend in the danger zone, the better. Prepared food should be covered and refrigerated within two hours. Try to divide your leftovers into shallow containers so that when you do refrigerate them they will cool quickly. Periodically stir soups and stews while they are refrigerated to help release the heat from the center of the storage vessel. Popping your leftovers right into the refrigerator while they are still hot is a much better alternative to letting them cool on a countertop at room temperature, but it is still not the best solution. The best way to cool any food product is in an ice bath. The more quickly the temperature drops to 40°F or under, the better.

Food Safety Tips (continued)

Defrosting Foods

Your mother may have told you that it is okay to thaw frozen meat on the counter all day while you are at work, but she could not have been more wrong. As I said earlier, freezing does not kill most bacteria, it simply slows them down. Once frozen food defrosts enough to reach temperatures in the danger zone, the bacteria will begin to multiply again. Just because a roast is frozen in the middle does not mean that the outer layer is also frozen. If it is being thawed at room temperature, it will spend a long time in the danger zone before the center of the roast is thawed. Likewise, thawing foods under hot water will also encourage bacterial growth. The best and safest way to defrost any food is overnight in the refrigerator. It is also safe to thaw foods under cold running water with constant drainage, though this method is wasteful and requires fairly constantly attention. Microwave thawing is also acceptable provided you promptly cook or refrigerate the thawed product. If you intend to refreeze a microwave-thawed product, cook it first.

Keep It Clean

One sure way to transfer pathogens from one food to another is through cross-contamination due to an unclean work surface. If you cut meat on a counter top or a cutting board, be sure to wipe it down with a disinfecting solution—bleach water works well—before beginning work on something else. Keep items that are not going to be cooked, such as salad greens, fruits, or other fresh vegetables, on a separate work surface. Finally, wash your hands frequently.

The Most Common Food-borne Illnesses

Six types of bacteria are responsible for most of the reported cases of food-borne illness:

Salmonellosis is the most common bacterial food-borne illness. The salmonella family includes about two thousand different strains of bacteria, but just ten strains cause most reported salmonella infections. Salmonella can be found on, and ingested from, a variety of sources, but is most frequently associated with poultry, meat, eggs, and unpasteurized milk.

Botulism is rare but very dangerous. It is usually associated with low-acid canned foods, such as meats and vegetables that have been improperly processed or stored. Botulism may also be present in dented cans, which should never be used.

Staphylococcal (staph) bacteria are present in humans in the nose and throat as well as in skin infections. Staph toxins are not killed by ordinary cooking, which is why personal hygiene is so important in the kitchen.

The Most Common Food-borne Illnesses (continued)

Clostridium perfringens are environmental, anaerorbic bacteria. Foods in buffets and casseroles, stews, and gravies, must be maintained at 140°F or above, or they will be particularly susceptible to growth of this bacterium.

E. coli 0157:H7. Recently attention has been focused on the tragic loss of life caused by *E. coli 0157:H7*, a rare strain of *E. coli* that produces large quantities of a potent toxin that causes severe damage to the lining of the intestine. The disease produced by it is called hemorrhagic colitis. *E. coli* 0157:H7 survive refrigerator and freezer temperatures. Once they get in food, they can multiply very slowly at temperatures as low as 44°F. The actual infectious level is unknown, but most scientists believe it takes only a tiny amount of this strain of *E. coli* to cause serious illness and even death. It is killed by thorough cooking.

Listeria monocytogenes is found in soil and water. Vegetables can become contaminated from the soil or from manure used as fertilizer. Animals may carry the bacterium without appearing ill and can contaminate foods of animal origin such as meats and dairy products. The bacterium has been found in a variety of raw foods, such as uncooked meats and vegetables, as well as in processed foods that become contaminated after processing, such as soft cheeses and cold cuts. Listeria is killed by pasteurization, and heating procedures used to prepare ready-to-eat processed meats should be sufficient to kill the bacterium; however, unless good manufacturing practices are followed, contamination can occur after processing. Since early August 1998, forty illnesses caused by a single strain of *Listeria monocytogenes* have been identified in ten states.

WHAT WE DON'T KNOW WILL HURT US

Finding our way back to a connection with our food is much more than fond memories and nostalgia—it may actually be a way of protecting our health. As we lose our food heritage we also lose age-old cooking methods and techniques, some of which may have been instrumental in making certain foods safe for consumption. Oral histories and folk tales once gave many people insight into their food sources. When I spoke to Loretta Barrett-Oden, a chef and an expert on Native American cuisine, she told me that food stories abound in the Native American culture. Manioc, often referred to as cassava or yucca, is poisonous, but indigenous peoples learned over time that the poison, naturally occurring cyanide, can be destroyed in part by heat in conjunction with specific preparation techniques. The process that manioc is put through, which was discovered and tested over time, consists of grating the cassava and draining it overnight prior to cooking. During that time much of the cyanide is released, and heat destroys the rest of the poison. Unfortunately, not all cooks understand the reasons behind it and have occasionally thought they had a better cooking method, like roasting, which might have proved fatal.

As processed foods become the norm, historical wisdom, caution, and everyday common sense are often lost in the rush to market. Nichols Fox gives an example of how ignorance of chemical reactions in foods can create incredible health risks:

> Responding to consumer demand for low calorie products, a British manufacturer of hazelnut yogurt decided to use a sugar substitute in the product. The result was an outbreak of *C. botulinum* that made twenty-seven people seriously ill and killed one. What happened? The simple and apparently innocent shift from sugar to a sugar substitute had altered the water activity of the product so as to allow spores of *C. botulinum* to germinate and produce toxin; with sugar, their growth had been inhibited.[49]

Mistakes like this one are not uncommon. To make matters worse, we have come to believe so blindly in the authority of our food processors and suppliers that we often trust them more than our own senses. Fox gives us another example: "In a large outbreak of foodborne disease from contaminated chocolate milk, many of those who drank the milk reported that it tasted funny but they drank it anyway. We trust our foods beyond reason, and that is a new development in history."[50] Without a doubt, this is a contemporary issue. Humans have a built-in gag reflex that automatically warns the body against bitter, poisonous foods, but for some reason we are training ourselves to ignore the very thing that has enabled humans not only to survive, but to evolve over tens of thousands of years.

Lest you think that food traditions are too nostalgic to be of use to you, whether you are a home cook or a chef, I will share a story that illustrates just how important our culinary traditions are. As I was reading Fox's *Spoiled: Why Our Food Is Making Us Sick and What We Can Do about It*, I came across a story about a New York restaurant that, in today's culinary fashion, is cooking its vegetables al dente. One particularly noteworthy evening was the one during which fiddlehead ferns, a New England delicacy, appeared on the menu. Of the guests who consumed the vegetable that night, three-quarters fell ill. Fox takes up the story: "What was new here was that people had begun preparing fiddleheads in nontraditional ways without a thought as to whether there was something protective in the traditional way of preparing them."[51] Traditionally, fiddlehead ferns are boiled for ten minutes before serving, a method that apparently rids the vegetable of a toxin capable of making a person rather ill. Unfortunately the chefs of that New York restaurant had no idea fiddlehead ferns could be poisonous. In fact, I did not know this myself until I read Fox's story. I was aghast at the realization that it could just as easily have been my cooks, under my direction, that had prepared the poisonous produce. Each year, for a short time in the spring, we serve fiddleheads. We also prefer to serve our vegetables al dente. Fortunately, even though I never knew the vegetable could be poisonous, I always boiled fiddleheads because that is the way I was taught to prepare them. I never questioned this, and was never told why. I subsequently took some time to look through twelve contemporary cookbooks, including those from a number of culinary schools as well as some written by America's greatest chefs. Only two of the books had information on fiddleheads, and only one of those mentioned boiling— as an alternative to steaming or sautéing. Nowhere was it mentioned that the vegetable is poisonous if not cooked correctly. I have spent nearly two-thirds of my life in kitchens, and this is a lesson I will never forget.

UMAMI: THE STORY OF THE PEACH

One of the reasons I am writing this book is that as a chef I am devastated that we have lost so much of the flavor in our foods through processing and mass production. I believe also that our health is directly attributable to flavorful, nutrition-packed foods. If we eat fresh local and regional whole foods, we will be reintroducing not only flavors to our lives but a healthier diet as well.

The Japanese have a term to describe food, *umami*, which in essence means not just the flavor of an item that is eaten at perfection but the experience of eating the items at their peaks. For me the ideal is captured by David Mas Masumoto in *Epitaph for a Peach: Four Seasons on My Family Farm*. He tells the story of his efforts to save Sun Crest peaches, which he believes to be the best eating peach in the world. Visually, Sun Crests are the "wrong" color, and most shippers and packers do not want to handle them because they are too soft. But when you read David's description, you want to reach through the pages and grab one—to me, that is what the Japanese mean by *umami*.

> The Sun Crest tastes like a peach is supposed to. As with many of the older varieties, the flesh is so juicy that it oozes down your chin. The nectar explodes in your mouth and the fragrance enchants your nose, a natural perfume that can never be captured. Sun Crest is one of the last remaining truly juicy peaches. When you wash that treasure under a stream of cooling water, your fingertips instinctively search for the gushy side of the fruit. Your mouth waters in anticipation. You lean over the sink to make sure you do not drip on yourself. Then you sink your teeth into the flesh, and juice trickles down your cheeks and dangles on your chin. This is a real bite, a primal act, a magical sensory celebration announcing that summer has arrived.[52]

The ideal that David describes is one that we have lost, the passion for wonderfully ripe seasonal food. Our communal attraction to having any product anytime we want it, our belief that our financial security can buy us whatever we desire, has placed our health and that of our children and families in jeopardy. Those perfect peaches, tomatoes, strawberries, and other foods consumed at the height of their seasons are unequaled in flavor. At their peak, they provide us with intense pleasure, yet we have willingly traded them for low-quality out-of-season products laced with chemicals and disease. It is our personal responsibility to understand the implications of our choices, to become informed about what we eat and feed our families, and then to demand that our food producers supply us with safe, clean, and healthful products. I hope, as we move on to the final chapter of this book, that you find yourself beginning to understand these issues, and starting to formulate the questions you should be asking. You have real power to change the way things work—beginning today.

·6·

The Future

There are three principal means of acquiring knowledge available
to us: observation of nature, reflection, and experimentation.
Observation collects facts; reflection combines them; experimentation verifies the result of that combination. Our observation of
nature must be diligent, our reflection profound, and our experiments exact. We rarely see these three means combined; and for
this reason, creative geniuses are not common.

—Denis Diderot, "On the Interpretation of Nature"

We all have food memories. The slightest aroma or taste of a special food
may conjure memories of family holidays, a first date, early childhood, or
even a treasured relative. One of my fondest memories is of my grandfather's tomatoes. He and my grandmother lived in Brookline, Massachusetts,
just outside Boston, in what can only be considered a city. They planted
tomatoes up the side of their yard, and every summer we eagerly picked
the sweet, bright red fruits from their vines and devoured them the way
one would eat an apple fresh from the tree. I want to make sure that my
nieces and nephews and their children have the opportunity to experience
the flavors of my childhood. To preserve the verity of our food supply, we
must find ways to support those food producers who are working toward
sustainability, or all we will have to bequeath to future generations will be
mass-produced food severely lacking in soul.

We literally hold the future global food supply in our hands, both com-
munally and individually, and there are some exciting alternatives ahead.
Our choices, ideals, and actions will shape the future. The food we buy, the
stores we frequent, the farmers who grow our food, and the organizations
and government leaders we support will dictate our future quality of life.
Now, at the beginning of the twenty-first century, we must work to educate our friends and families and begin down the road to a sustainable
tomorrow.

POPULATION

The world's population will more than double in the next twenty-five
years. This dramatic increase will impact the global food supply in ways

not yet entirely understood. The biggest increases in population will occur in the Third World and developing nations, due primarily to religious conviction and lack of family planning education, which makes it easy for us here in the United States to focus blame there for overpopulation and overconsumption. However, we would do better to turn our gaze homeward. In fact we Americans have about forty times the impact on the food supply as those in developing countries. Paul Hawken reports that "the 50 million people who will be added to the U.S. population over the next forty years will have the same global impact in terms of resource consumption as 2 billion people in India."[1] Those figures speak volumes about the unchecked abundance we enjoy in the United States. They also reflect a high level of dispassionate irresponsibility among our citizenship.

WHAT WE CAN DO

Agriculture

The importance of supporting community farmers is paramount; they keep local economies strong and help conserve fossil fuels. It is not enough, however, to seek out area farmers without considering their agricultural methods. If a local farmer is using large amounts of herbicides and pesticides, is monocropping on a large scale, or is wasting precious soil and water, you would do better to seek out a producer in a neighboring state who uses sound farming practices. Do not, however, mistake innovation and technological advance for poor farming practices. In fact, efficiency and technology are not necessarily contrary to sustainability. As Eliot Coleman points out, "Viable production technologies exist that address environmental and economic realities. Some of these production technologies may require new ways of thinking, while others may appear to revive old-fashioned or outmoded ideas."[2]

Coleman's own example of New England winter farming is one form of agriculture that should be looked at as a model for the future. His cold frames and plastic tunnels could dramatically improve the diversity of regional food webs. John Jeavons, like Coleman, has put a great deal of effort into developing alternative agricultural techniques. Through biodynamic intensive farming, Jeavons works to produce the optimum yield in the smallest possible space, continually producing amazing results. The per acre production on land farmed biodynamically is four to six times greater than the average U.S. yield, and it is all done without pesticides or chemical fertilizers. Another method we should be supporting is integrated pest management, which concentrates on utilizing plants and animals to holistically control detrimental insects. For example, a recent Associated Press article reported that pecan growers could help protect their trees from stinkbugs by planting peas around their orchards. The bugs are enticed by the peas, which results in dramatically reduced infestations in pecan orchards.

An excellent way to become directly involved in local farming is to support CSAs. The money you invest at the beginning of the season will go directly to meet the costs of running the farm. In addition, if you are interested, most CSAs will give you a discounted rate in exchange for part-time labor. Every week during the harvest period shareholders receive a box of fresh, organic vegetables and/or fruits. CSAs keep money in the community and save on fossil fuel use, but they also help us to understand seasonality and teach us the difficulties and rewards of farming sustainably in ways that cannot be accomplished by other means. When a particular crop is ravaged by bad weather and another produces far beyond expectation, shareholders understand how it happened because they were part of the process. The real joy of farming this way is in the incredible flavors your food will have; the farmer is not limited to products that ship well or can withstand modern processing or packaging techniques.

Support and Demand Diversity

We absolutely must support diversity in all of our food supplies, not only for the preservation of flavor and food memories but for long-term food security. There are many ways to do this—supporting local farms, shopping at farmers' markets, buying and planting your own heirloom seeds, and making a conscious effort to purchase animal products from sources that use more obscure breeds than commercial agribusiness farms do. Part of your commitment to biodiversity has to be a willingness to spend more money on the food you buy.

Redirect Corporate Dollars

It is naive to think that all of the world's food can be grown on small, local farms, so we have to find ways to include agribusiness in our sustainable future. A recent report by PR Newswire is evidence that at least some large agribusiness companies believe that sustainability and profitability go hand in hand. Amway Corporation, made famous by its line of consumer direct sale cleaning products, has acquired a majority interest in Trout Lake Farm, the largest organic herb farm in the United States. The farm will provide Amway with organic, high-quality herbs for use in its food supplements. Such a keen interest by a corporation in an organic farm confirms the belief that consumer demand dictates marketplace cycles.

Memories of a Community Supported Farmer

Linda Halley, Harmony Valley Farm

The first car parks across the road, a blue minivan. It pulls up next to the FMC beet digger, a hulk of iron that looks more like a dinosaur than a farm implement. A couple gets out with a picnic basket and a stroller. I can see them eyeing the digger suspiciously. Out pops the stroller-rider, heading straight for the iron dinosaur. She's snatched back just in time; no cuts, no bruises, no dirt, and no grease. Why did we invite these people to invade our farm? I don't know them and they are obviously foreigners to the world of agriculture. I trot down the driveway with a welcoming smile on my face anyway. "Hi, glad you could make it! I'm Linda," I muster. "John, Leslie, our daughter, Analise." We shake hands. Turning to their daughter, buckling her into the stroller, "This is Linda. She grows those yummy strawberries that come in the box." They look up, genuinely happy to be here, "Oh, this place is so beautiful!" "Thanks, we like it. Come on up to the house." I can feel a little warm happiness beginning in me as well. John struggles, pushing plastic wheels through the clumpy clover and then up the driveway. Before we get far, another car arrives, "Go ahead," I say. "Make yourselves at home. I'll be right up, and Richard's up there somewhere, too." John hauls his smiling child over the 2-inch gravel designed, obviously, with tractor traffic in mind. The road is suddenly busy with cars, blinkers signaling a turn into our field. Folks to greet, hands to shake, children to help. . . . I'm feeling a bit overwhelmed and unsure of whether this was such a good idea. I glance toward the house. Analise has been set free, the stroller wisely abandoned, and John and Leslie are poking their heads into the greenhouse. I nod my head involuntarily. Maybe it's not going to be such a bad day after all.

That was seven summers ago. It was the morning of our first CSA event—a potluck and field tour. Everyone learned a lot that day—especially this farmer. People are hungry in many ways. Not everyone thinks about food in the same way. It feels good to be appreciated for what you do. A farm is not a park. Growing food is a lot of work. You can feed a lot of people off one acre, even more off forty! When food is prepared with love and respect, it tastes extra good. When food is grown with love and respect, it tastes extra good. Contrary to popular belief, kids love vegetables.

Richard, my husband and farming partner, and I continue to learn a lot from our CSA members, and they in turn learn from us. That's part of what makes being a CSA farmer, or member, different than just growing, or buying, produce in the conventional way. In an atmosphere of "get big or get out" agriculture, we are expected to compete on price with thousand-acre corporate farms half a continent away. When selling to distributors we are regularly quoted prices "Freight On Board—California" and sometimes given a thin dollar-per-box premium as a concession to freshness. It has become clear that if our farm is to thrive, we must educate ourselves and the buyers of our produce. CSA gives us the opportunity to do that. Or, perhaps I should say, *challenges* us to do that!

From a farming standpoint, CSA members demand a lot! They want variety, quality, flavor! "Standards" like carrots and broccoli have to be available often and over a long season. Unique crops like sugar snap peas, lacinato kale, or ground-cherries add necessary variety. Outstanding quality is a must for seasonal favorites like sweet corn and tomatoes. And don't forget everyone loves berries! That's not expecting too much, is it? Perhaps not. That's what it takes to keep savvy eaters of the 1990s happy. But let's face it, being a good grower of forty different crops over a long season, May through December, no matter what the weather, would challenge even the most expert farmer!

Memories (continued)

Add to that the complications of dealing with many (hundreds in our case) of individual households—receiving their phone calls, explaining the delivery system, collecting their payments, and maybe even giving suggestions about how to serve fennel at their family reunion. Compared to the alternative, a handful of wholesale buyers with fax machines, a farmer begins to ask, "Why?" To answer that, Richard and I refer back to that first CSA event. The day began with a potluck feast unlike any other I had known. Members had transformed the vegetables we had just delivered to them the day before into scrumptious, nourishing delights. They obviously chopped, sliced, and simmered with an appreciation for the plant it was and the food it could become. Later we toured the farm aboard the canopied harvest wagon pulled by the proud 1950s-vintage Allis Chalmers tractor. We stopped frequently, letting the kids scramble off the wagon and out into the fields. Such wonder for the little ones to peer inside the long, tied leaves of a cauliflower and spy the blanched curd, like a hidden jewel. We talked at length about our farming system—why we plant hedgerows of willow, what "green manure" is, how we keep the rows weed-free, and how that iron contraption actually digs beets.

The final stop was the strawberry patch. No more need to talk. We all fanned out with a little container in one hand and our eyes on the red prize. Sweet, sun-warmed, juicy. Our fingers were stained red—our lips, too. No one needed to tell the children which ones were ready to pick—taste alone was their teacher. By the end of the day Richard and I felt like actors at a curtain call. Our efforts were recognized. Our guests, the CSA members who had become our friends, had a new appreciation for where and how their food was grown and their own importance in the equation. They, too, felt recognized for their support of the farm, monetarily and emotionally. Since that first CSA event we have hosted many more. We've come to cherish the relationships forged out in the fields with our members and their families. Many of the youngest have grown up on fresh food mostly from our farm. Some members send cards in December, with holiday menus revolving around our winter produce proudly penciled on the back. Some enclose pictures, "Isabel's first solid food, Harmony Valley Farm squash." But you can't trade fond memories for all the work and investment it takes to run a farm. We get more than that from our CSA members. They support us economically by buying for the season. It reduces that roller coaster ride that can come with a farmers' market business, where a rainy day can mean most of a harvest going unsold. They become contributors to their local economy—returning money directly to the producer rather than dividing it up to pay for long-distance trucking, refrigerated storage, advertising, or commissions.

Through our weekly newsletter our members become aware of food production issues and economic, environmental, and health considerations. It gives them a farmer's perspective to consider when making food purchases. CSA members buy from us knowing that "FOB California" prices don't really reflect what good food costs to grow. And let's not downplay the value of a long-term relationship. John and Leslie are still members. They opened their garage every Saturday as a delivery site for six years! They've been known to claim only a fuzzy recollection of what their local grocery's produce aisle looks like. We've come to the conclusion that there are easier ways to make a living growing produce, but there is no more rewarding way than through community supported agriculture.

Influence Allocation of Governmental Dollars

We should begin by learning to work with the government to control the power of agribusiness and direct its research in ways that encourage sustainability. Subsidies whose primary beneficiaries today are large corporations are contrary to the spirit in which they were originally conceived. Farm supports ought to be directed away from businesses—like corporate hog factory-farms, or even smaller conventional farms—that have a negative environmental impact and given specifically to those that display marked interest in preserving the environment. Rather than encouraging corporate farming to continue unchecked, this type of farm support would give corporate farmers an incentive to make moves toward sustainability.

One good example of this idea can be seen in an award presented in 1999 by the President's Council on Sustainability. Bates College Food Services in Lewiston, Maine, received the council's award for feeding a hundred homeless people daily, providing a stable market for local organic farmers, and making pigs at the Ricker Farm happy, all in one fell swoop. At Bates College, they have accomplished this through innovations in providing meals for the university's students, faculty, and guests. Through a co-op of local farmers, the college buys seasonal, organic food, providing farmers economic stability. A local farmer composts the college's preconsumer food waste, and pigs enjoy the food scraps. Uneaten cafeteria food is distributed to a local soup kitchen. These efforts have forged a powerful connection between the college and the community, reduced transportation and pesticide impacts, lowered disposal fees, and provided healthy, high-quality food to students and faculty. This is the kind of creative idea that our tax dollars should be subsidizing. In this case the award was not a monetary one, but perhaps in the future it should be. Write your congressmen, let them know that you support these programs, and send e-mail, letters, faxes, and use your voice to show support. It truly matters.

KEEP AN OPEN MIND

It is easy to be swept into a way of thinking that does not allow unique or innovative approaches to enhance or expand the current ideology, so it is extremely important that each one of us remain open to all possibilities. Even biotechnology must be seen as a potential source of sustainable food for the future. The Rockefeller Foundation, a private philanthropic organization, for example, aims to "increase crop yields of small-holder farmers in developing countries profitably and without degrading natural resources," and in 1984 the foundation began the International Rice Biotechnology Program, focusing on Asia. The program is an integrated set of research, training, technology transfer, and capacity-building activities that will benefit low-income rice producers and consumers in developing countries. The foundation's hope is to increase rice production in Asia by 20 percent by the year 2005 through the use of biotechnology without damaging the environment or reducing farm incomes. While I still do not feel in a position to either denounce or advocate biotechnology, I do believe that the industry may hold the key to some types of sustainable food production, providing we work to guarantee the long-term future safety of our food supplies.

FISHERIES

We must work to ensure the sustainable viability of our seafood supplies through support of aquaculture research and technological advance. A good example is in recent research that has shown catfish, long a stalwart of Louisiana cookery, to be particularly suited to farm raising. Catfish are native to the Southeast, making the potential for species pollution relatively small, and what makes them particularly attractive as a farmed fish is that they are at the bottom of the food chain. Largely herbivorous, catfish require only about 3 percent fish meal to thrive. Shellfish, most notably mollusks, have also been shown to be well suited for aquaculture because they are natural water filters. Mollusks feed by filtering small particles of plankton and algae from the water. They grow all around the country in basin estuaries and actually clean water rather than pollute it. They do produce a feces-like by-product, but their overall effect on the environment is much more positive than negative.

Polyculture, long practiced in China and Japan, is aquaculture at its best. As biologist Rebecca Goldberg explains, polyculture is the farming of many species of plants and animals together in one system

> in order to make optimum use of the water and nutrients and to minimize farm wastes. Aquaculture systems that produce hydroponic vegetables with fish appear to be increasingly common in the United States. These systems grow crop plants with aquatic manure, suspending the roots of crops in aquaculture effluent. These crop plants remove large quantities of nutrients from effluent in order to nourish their growth, thus cleaning the effluent. Sale of these vegetable crops then generates income for aquaculturists. The Inslee Farm, Inc. of Oklahoma, for example, grows chives in greenhouses using the effluent from ponds in which a variety of different fish species are raised, including tilapia, catfish, and grass carp.[3]

Some polyculturists have catfish growing on the bottom of the tanks, finfish in the next layer above, mollusks on ropes growing vertically, and hydroponic vegetables on the uppermost layer. Polyculture can also be part of a terrestrial agriculture system in which the nutrient pollution (wastewater by-product of aquaculture) is fed to poultry and used as fertilizer on fields.

Technological advance in existing aquaculture systems is extremely important. Dick Monroe asserts that fish husbandry must evolve just as animal husbandry has: "I believe that our ancestors will look back on the millennium that we are about to approach and see it as the cusp between still hunting sea creatures, and husbanding them. With the knowledge that is coming through rather quickly in aquaculture of lesser-known species, we could see not only trout, catfish, salmon and tilapia being raised, but grouper, striped bass, redfish, mahi-mahi and tuna."[4] Jim Carlberg of Kent Seafarms, who is already farming striped bass, believes that some of his success is related to water temperatures that remain constantly warm due to the "hot springs" in the surrounding area. Jim explains that "by having a constant elevated temperature of approximately 78°, we are able to grow fish from essentially a small fingerling that is about a month old, to market size, in as little as ten months."[5] Perhaps more locations that have natural warm water should explore aquaculture as a farming alternative.

We also need to support research and technological advance for wild fish harvesting. At an Oldways symposium on sustainable food supplies, better methods of fish farming were discussed: "Of all the sea's possibilities for man's future, the greatest may be its promise of a significant increase in the world's food supply. Square mile for square mile, the sea is estimated to be potentially more productive than the land. Yet at present the oceans supply only 1 or 2 percent of man's food. . . . For the future world population man will have to start large-scale farming of the sea as he has for so many thousands of years farmed the land. If such scientific techniques are perfected and extended, aquaculture—farming of the sea—may one day provide the world with a constant supply of the protein so desperately needed by burgeoning populations."[6] One test reported at the symposium showed that a one-acre shellfish farm can yield around 15,000 pounds of protein, a figure that could never be equaled in land-based farming.

Consultant David Wills told me that Darden Restaurants is working on an effort to make wild fishing more sustainable. One of his favorite programs is restocking: "I like the idea of growing captive strains of stock and putting them back in the ocean so they can be recruited into the wild cult systems. Darden Environmental Trust, which is the environmental arm of Red Lobster, has been building artificial lobster 'condos' in different parts of the world and sinking them into the ocean where reefs have been bleached out, or other toxicity factors have destroyed the reef. You seed them with a couple young adult lobsters, and you come back in a year, and every little crevice and cranny of say a forty- to fifty-cinder-block condominium is filled up with lobsters who use it for protection and safety."[7]

Making the oceans a sustainable resource will require us to make more difficult decisions than where to put lobster condos, however. Conservation limits will have to be implemented and enforced on a global scale, and preservation efforts must be undertaken before a species like Pacific salmon is so overfished that it is placed on the endangered species list. The fishes' numbers, in some cases, are a mere tenth of what they were just a century ago. If loggers, sport and commercial fishermen, grocery store owners, and restaurateurs had only taken steps to preserve the species earlier, their livelihoods might not be in jeopardy today. It is also imperative that consumers embrace businesses that support environmentally safe harvesting and production practices, and discontinue support of those that do not. Consumer boycotts have been effective tools in changing fishing practices in the past, most notably a tuna boycott brought about by public sentiment against unnecessary dolphin losses due to irresponsible fishing practices. Ultimately, if you care about preserving our fisheries, it is your responsibility to buy seafood that was raised and/or caught sustainably. Before you buy fish, ask where it came from. If it was farm raised, do some research on the company that brought the fish to market before handing them your hard-earned dollars. Demand that policy makers support scientific recommendations on fisheries management, and whenever possible eat local fish, support local fishermen, and try new menu items to promote biodiversity. Even what seems like the smallest effort will make a difference.

BECOMING FARMERS

Mohandas Gandhi once said, "To forget how to dig the earth and tend the soil is to forget ourselves." One way we can positively impact our food supply is to become farmers, on a small scale. For some this might mean herbs in the kitchen, for others, a more extensive kitchen garden, and still others might be able to handle an acre out back. In whatever way you are able, try it. Grow some food and share the experience with your friends and family. Edible landscaping, the artistic use of food plants and trees, is one way to begin "farming" around your house, school, or place of work. Rather than planting the usual pansies, daffodils, and tulips, plant herbs and edible flowers. Instead of another pine or oak tree, give the space to a fruit-bearing tree. Start a little tomato garden. You will be surprised at how much food you can grow in a small area.

In fact, in just a small backyard you can grow much of your own food and provide more than adequate sustenance for your family. Joan Gussow spoke with *Health* magazine about her own backyard farm at her home in upstate New York, which she tends with her husband:

> "We are trying, as much as possible," she says, "to eat what the garden can provide." Inspired, they laid brick walkways and built raised beds, double-

digging the soil, adding compost, and figuring out which vegetables to plant side by side to ward off insects and diseases. The true health benefit of home-grown produce, says Gussow, is that you will eat more of it, period. "There is just no comparison between vegetables you get in the store and ones you pick from a garden. Real vegetables are so intensely flavored when they are fresh that you do not need all the salt, sugar, and fat that are loaded into our food. When you get used to the flavors of fresh vegetables, you do not want anything else. I do not like to eat things out of season. They do not taste right. Canned asparagus? Winter tomatoes? Forget it. It is so much better to build up the anticipation of that first fat blueberry or the first red pepper when the season finally comes."[8]

These days it is easy for the average person to get an incredible variety of seeds. The Cook's Garden, Johnny's Selected Seeds, and Burpee Seeds have all seen tremendous growth and are striving to take advantage of the country's renewed interest in vegetable gardening, especially organic vegetable gardening. Johnny's operates its own certified organic trial and research farm, where it evaluates over 4,000 new varieties a year. Cook's Garden has its organic gardens in the Intervale in Vermont, and Burpee has seen its business grow to include 500 different vegetable and herb varieties. All three companies have websites with online shopping capabilities. Ordering their catalogs or shopping online is a great way to learn about different products that will grow beautifully in your part of the country.

HEALTH CARE COSTS

Heart disease, cancer, obesity, and high blood pressure are all directly attributable to poor eating habits, and each and every one of us is feeling it in our pocketbooks. Our eating habits are a major contributor to the rising cost of health care in America. Irena Chalmers discussed this subject at an Oldways conference on sustainable cuisine: "Current American health care costs are 120 times greater than those in Japan, where the diet is lower in cholesterol. We cannot remain competitive in a global economy when we devote so much of our resources to medical costs. For example, an American car costs $500 more than a Japanese car, and that sum is directly attributable to expenditures for the health care premiums for the workers."[9] National health care costs have topped $1 trillion, to which each American contributes approximately $3,700 annually. It has been estimated that about one-third of the cost of American health care could be eliminated by increasing exercise and eating a nutritious diet. If we could redirect, by changing our eating habits, even 10 percent of the monies spent on health care to benefit sustainability research, imagine how much closer we could be to our goal of a sustainable food supply.

Produce Seasonality Chart

Fruit or Vegetable	Jan.	Feb.	Mar.	Apr.	May	June	July	Aug.	Sept.	Oct.	Nov.	Dec.
Apples	□								□	□	□	□
Apricots					O	●	●	O				
Artichokes			O	O	O	O						
Arugula	O	O	O	O	O□	O□	O□	O□	O□	O	O	O
Asparagus			O	O■	O■	O□						
Avocados	O	O	O	O	O	O	O	O				O
Beans, shell			O□	O□	O□	O□	O□	O□	O□			
Beans, snap					O□	O□	O□	O□	O□	O□		
Beets, gold							O	O	O	O□	O□	O□
Beets, red						●	●	●	●	O□	O□	O□
Berries						O	O	O	O	O		
Blood oranges	O	O	O									
Bok choy									□	□		
Boysenberries						□	□	□				
Broccoli						O□	O□	O□	O□	O□	O□	O□
Broccoli rabe	O	O	O	O			□	□	□	□	□	O□
Brussels sprouts	O	O	O							O□	O□	O
Bulb fennel	O	O	O	O	O				□		O	O
Cabbages	O	O	O	O	O	O	O	O	O	O	O	O
Cantaloupes						O	O□	O□	O□	O		
Carrots	O	O	O	O	O	O	O	O	O	O	O	O
Cauliflower						O	O	O	O	O		
Celeriac	O	O	O	O	O	O	O	O	O	O	O	O
Chard	O	O	O	O	O	O□	O□	O□	O□	O□	O	O
Chayote	●	●									●	●
Cherries						□	□					
Chinese cabbage										O	O	
Collards								O□	O□	O□	O□	
Corn						□	■	■	■	□		
Cucumbers	O	O	O	O	●	●■	●■	O	O	O	O	O
Eggplant, baby				O	O	O□	O□	O□	O□	O		
Eggplant, white						O□	O□	O□				
Endive								O	O	O	O	
Escarole	O	O	O	O	O□	□	□	□	□	O□	O	O
Fiddlehead ferns		□	□	□	□							
Figs							O	O	O	O		
Frisée	O	O	O	O	O	O	O	O	O	O	O□	O□
Gooseberries								□				
Gourds									□	□	□	□
Grapes						O	O	O	O	O	O	O
Herbs			O	O□	O□	O□	O□	O□	O□	O	O	O
Jerusalem artichokes	O	O	O	O	O						O	O
Jicama	O	O	O	O	O	O	O	O	O	O	O	O
Kale								□	□	□	□	
Kiwi	O	O	O			O	O	O	O	O	O	O
Kohlrabi						●□	●□	O□				
Kumquats	O	O	O									

Fruit or Vegetable	Jan.	Feb.	Mar.	Apr.	May	June	July	Aug.	Sept.	Oct.	Nov.	Dec.
Leeks										O	O	O
Leeks, baby						O	O	O	O	O	O	O
Lettuce, baby			O	O	O	O	O	O	O	O	O	
Lettuce, bibb	O	O	O	O	O	O□	O□	O□	O	O	O	O
Lettuce, iceberg	O	O	O	O	O	O	O	O	O	O	O	O
Lettuce, leaf	O	O	O	O	O	O	O	O	O	O	O	O
Maché	O	O	O□	On	On	O□	O□	O□	O	O	O	O
Mangoes	O	O	O	O	O	O	O	O				
Mustard greens							O□	O□	O□			
Nectarines						O	●	●	O			
Okra							O	O	O	O		
Papayas	O	●	●	●	O	O	O	O	O	O	O	O
Parsnips				O	O	O	O	O□	O□	O□	O	O
Peaches							□	■	□			
Pears	O□							■	■	■	O□	O□
Peas, green	O	O	O	O	O	O	O					
Peas, snow	O	O	O	O	O	O	O					
Peppers, green bell						O	O□	O□	O□	O	O	O
Peppers, red bell								O□	O□	O□		
Pineapples	O	O	●	●	●	●	●	●	O		O	O
Plums						□	■	□				
Potatoes, baby red					O	O□	O□	O□	O□	O		
Radicchio	O	O	O	O		□	□	□	□	O□	O	O
Raspberries						□	□	□	□	□		
Romaine	O	O	O	O■	O■	O□	O□	O□	O	O	O	O
Rutabagas									□	□	□	□
Salsify	O	O	O	O						O	O	O
Scallions								□	□			
Spinach	O	O	O	O	O■	O■	O□	O□	O	O	O	O
Squash, acorn									□	□	□	□
Squash, baby							O□	O□	O□	O□	O□	
Squash, butternut									□	□	□	□
Squash, golden									□	□	□	□
Squash, pattypan							O□	O□	O□	O□		
Squash, spaghetti									□	□	□	□
Squash, zucchini				O	O	O	O□	O□	O□	O□	O	
Strawberries					□	□	□	□	□			
Tangerines	O	O	O									
Tomatoes						O□	O□	O□	O□	O□	O□	
Tomatoes, cherry							O□	O□	O□	O□	O□	
Turnips	O	O	O	O	O	O	O	O	O□	O□	O□	O
Watercress	O	O	●	●	O	O	O	O	O	O	O	O
Watermelon					O	O	O	O	□			

California: O   Northeastern U.S.: □
California Peak: ●   Northeastern U.S. Peak: ■
(Chart taken from *The New Professional Chef*, 6th ed.)

SO WHAT IF WE ATE LESS MEAT?

Reducing our consumption of animal protein is not only healthier for our bodies but better for the earth. In *Vital Signs 1998: The Environmental Trends That Are Shaping Our Future*, Lester Brown explains,

> Perhaps the single most important distinguishing feature of dietary changes over the last half-century has been the growing appetite for animal protein. It is hunger for protein that spurred an increase in the world fish catch of nearly fivefold, boosting it from 19 million tons in 1950 to 93 million tons today.... Worldwide, the production of beef and mutton, like that of fish, depends heavily on a natural system—rangelands. And, like oceanic fisheries, rangelands are being pushed to the limits of their carrying capacity and beyond. Once rangelands are fully exploited and substantial growth in beef production can come only from feedlots, then the competition with pork and poultry for grain intensifies.[10]

As we already know, cattle and other livestock consume 90 percent of the soy and 70 percent of the grain grown in the United States. That soy and grain could be used to feed starving populations around the world, 40 to 60 million of whom die annually. The time has come to return to more traditional diets in which meat is not the focal point of the meal but simply a garnish—a treat and special accompaniment to a meal, not the meal itself. We must create meals with a focus on grains, legumes, and a variety of fruits and vegetables.

DEMAND LOCAL

Food in the United States travels an average of 1,300 miles prior to consumption. The farther our food supplies travel from their points of origin, the more they deteriorate, resulting in products with compromised nutritive value and flavor. Supporting farmers' markets and CSAs and demanding local products at our supermarkets are a few of the ways we can all cut down on overuse of fossil fuels and increase the nutritive value of our food.

In the past decade we have seen the number of CSAs in America increase to over 1,000 nationwide, serving a total of over 100,000 households. They vary in size from plots that sustain ten families to full-fledged farms with as many as 700 members. You might be surprised to find that there is one right in your neighborhood.

IT ALWAYS COMES BACK TO MONEY

Part of the challenge of working toward a sustainable future is financial. As you know, it is often less expensive to buy mass-produced, well-traveled food products than it is to buy local products. In the United States the idea of cheap food has become a sort of emblem, but cheap food is not a bargain, it is an illusion—a mortgaging of the earth so complete that future generations may not be able to pay our debt. What we "save" today at the grocery store may cost us many times more in environmental degradation and related health costs in the long run. The pollution, fossil fuels, water, garbage, and chemicals that make inexpensive food possible today will only cause harm in the future. Pay a little more now, and you may just help change the future.

CHEFS, RESTAURANTS, AND DEEP WEALTH

One of the challenges of all food service operations is controlling food cost. In many restaurants this often tips the scales toward buying industrially produced rather than local, regional, or organic foods. Every chef I spoke with addressed this issue, and all of them were working to find ways to alleviate the problem. One of the most common solutions is to pass the costs along to the consumer, but this will not work without dedicated effort on the part of the restaurateur to educate his clientele. At the Putney Inn we highlight local and regional ingredients by listing the names of our farmers and producers right on the menu. We also educate our waitstaff about what we do so they can educate the consumer at the point of sale.

Chef Nora Pouillon feels that one way she can keep costs in line is by inspiring her employees to have passion and respect for food: "Nothing is thrown away in my restaurant. When we clean vegetables, we keep the peels and stems and make vegetable stocks. The stems from broccoli are made into soup, and the orange peels are cooked and pureed and made into orange glaze. Everything is used. I feel that I am an example that you can run a business, have organic food, and have an innovative, creative, high-end sustainable restaurant in an elegant setting and make money."

(In 1999 Restaurant Nora became the country's first certified organic restaurant. It is my hope that Nora's accomplishment will help other restaurateurs to see organics as part of their restaurants' sustainable futures.)

Michael Romano says that invariably local produce costs more money and that he sees a little spike in food costs in the summer, but he has accepted it as part of the cost of doing business. "It is a decision that we have made, a choice that we make, and we just work with it. There is more to it than just the best price we can get, it is the feeling we get of doing our share to help sustainability, and to help support local agriculture."

Rick Bayless has had some creative ideas for working toward sustainability. For starters, a couple of years ago he began trying to lower food cost and invest more in labor cost.

> Without changing the focus of our buying, we decided that we were going
> to eliminate certain ingredients that we were getting that were *very* expen-

sive, and as well we decided that we were spending too much money on protein items. What we decided was that we were going to cut back on some
portion sizes, and that we wanted to spend much more of our dollars on food
from local farmers. We decided we were going to buy as little stuff from the
outside as possible, and when we did it would only be the cheap stuff that
we were buying from the outside. So that we could pay more money to our
local farmers, to support them and hopefully grow their businesses, so that
other people would want to do the same thing. We got zero negative response
from our customers when we cut the portion sizes, it was very exciting.[11]

Rick's ideas are reflective of what chef Odessa Piper refers to as "deep
wealth," which I believe is at the heart of sustainability. The idea is that
wealth ought to extend beyond money in the bank to include lifestyle, ideals,
and an investment in the future:

> I want to introduce the concept that sustainability and supporting local agri-
> culture is an investment in the future, and this is not immediate wealth, this
> is deep wealth. I started using this term recently, 'deep wealth,' and I think
> I really like it, because a lot of wealth has simply been defined very two-
> dimensionally, as you know, that bottom line 10 percent. But what about
> the wealth that gives you 5 percent, and the other 5 percent that you are
> not getting, is coming in the form of goodwill, longevity of staff, loyalty of
> customers, and stability of the farms that you do business with. In fact, per-
> haps five years from now, with all of those pluses going towards you, your
> business will be even more prosperous, because of those aspects.[12]

In Odessa's model, the almighty bottom line should not and cannot be the
driving force if the ultimate goal is a healthy and sustainable food supply.

CHILDREN AND EDUCATION

One of the most important aspects of deep wealth is the next generation.
To move sustainability forward, children must be well educated about their
choices for the future. In our rush to teach our kids to navigate the information superhighway, we are forgetting to teach them about the system of
life that surrounds and supports them. Early exposure to caring for the
earth will enable children to better nourish themselves and others. School
gardens are a necessary part of their education. They provide a living learning environment and teach kids basic math, agriculture, horticulture, and
cooking skills. We are seeing a resurgence in school gardens across the country, but their numbers are nowhere near those of a century ago, when there
were 80,000 school gardens in the United States. It is my hope that we will
again see school gardens in every town in America. In fact, this is not an
unattainable goal.

San Antonio, Texas, is a prime example of a city that has taken simple and dramatic steps toward creating school gardens. In 1990 the San Antonio Independent School District partnered with the Bexar County Master Gardeners Programs and committed to establish eight new gardens every fall in inner-city schools. Five years later over 10,000 students were actively participating in the program, and over 300 volunteers were contributing more than 21,000 hours of community service annually. At the time of this writing, nearly 200 San Antonio schools have created a gardening program with the help of the Texas Agricultural Extension Service at Texas A&M.

In Vermont, a group of school garden advocates is working on reviving the teaching garden. Together they have authored a book titled *Digging Deeper*, the aim of which is "to cultivate the budding movement for practical, place-based education by providing a new generation of teachers, community workers, children, and parents with the fundamental tools to grow their own food safely and reliably. Together with thousands of communities around North America and the world, we see this as a crucial step toward genuine sustainability locally and human-scaled cooperation globally." The book is a real how-to guide to setting up school gardens. From a set of criteria to use in determining how to set up the garden to diagrams of actual gardens and recipes, this book is a wonderful resource, and can make starting a school garden an easy and rewarding endeavor.

Institutions of higher education are doing their part as well. Vern Grubinger, instructor at the University of Vermont, recently described some of the educational opportunities that exist at the college. Though the University of Vermont is not a culinary school, its foundation as a land-grant university gives it strong agricultural roots. More and more universities are introducing sustainable agriculture, or ecological design, programs as well as agro-ecosystem centers and student farms. Both ecological design and agro-ecosystem work in concert with the land as opposed to redesigning the land to fit farming. UVM's internship program is on the cutting edge of agricultural education in sustainability models. It sends students to all corners of the state to participate in every type of agriculture, including dairying, produce farming, animal husbandry, and specialties like cider and cheese making. Grubinger says, "One thing we are trying to do is connect students with real agriculture; we have a sustainable agriculture internship program which is a four-credit experience on a farm. Our internships require a learning contract, and they require some academic rigor, so the student is a colearner with the farmer and actually studies some aspect of the farm system related to sustainability. But it is basically in the simplest terms about sustainability, which means profitable stewardship of resources, and connection to the community." One internship project helped farmers study the feasibility of converting from conventional to organic dairying. Another examined the use of chickens as pest controls in a flower garden. Yet another helped a local cider producer develop a marketing plan. These internship programs not only supplement the students' educations but also assist farmers who may not have the resources to conduct their own research. UVM also sponsors the Land Link Vermont program, which works to connect young farmers with experienced agriculturalists who are seeking to transfer their land and experience to farmers who will be able to carry the torch into the future. This is particularly important because there is a shortage of young farmers in America. Vern explained that the school has built a database

> in a sophisticated way so that both sides list the attributes they are interested in. For instance, do I need bottom land or pasture land, do I want to be near a population center for retail or not, and then we give them the best matches. . . . It costs a lot of money to buy a farm in Vermont these days, so we are looking for opportunities where there is either a lease with option to buy, or a partnership, or an employee with a potential to buy in later. The exciting piece of this is, if you could plan ahead before you retire from farming for about ten years and find somebody, and actually work with them for five or ten years, you have not just transferred the land and the buildings, but all of your knowledge about the enterprise.[13]

The Culinary Institute of America has an organic garden and organic herb terraces at its Greystone Napa Valley campus, but there is no such garden at the original Hyde Park campus, where plans have been in the works for over five years. Executive Vice President Tim Ryan told me that gardens and a year-round greenhouse are part of the school's future, but funding is lacking. I feel strongly that growing food should be an integral part of professional culinary education. It is one way to help future food experts to truly understand food, seasonality, and sustainable food choices. Eve Felder, CIA assistant dean, dreams about the inclusion of these types of programs at the Hyde Park campus. "I would love for us to have in our curriculum, a program whereby our students . . . have to garden a 50-foot by 50-foot plot within the two-year period that they are a student here. They would have two years to do it, to plant seeds, grow them, develop recipes from what they have grown, and see the challenges that farmers face. I think traditionally we have been in antagonistic relationships with our purveyors. If you walk in someone else's shoes and know what it is like when you have a drought, and all of your squash keels over and dies, you can relate to a farmer in a major way and be more empathetic."[14]

Joe Baum was an icon in the restaurant world for over five decades; his restaurants included Windows on the World and the Rainbow Room, and in New York City he was on the cutting edge of creative restaurant entrepreneurship. In 1999 Joe Baum's legacy, the Joe Baum Forum on the Future, had its beginning with a sustainable cuisine conference at the CIA Hyde Park campus. I worked with Joe's daughter Hilary, Tim Ryan, and other CIA department heads to develop the project, which had as its goal to educate participants about ways in which sustainability can be addressed in the hospitality industry. When industry professionals and premier culinary educational institutions join to educate chefs on sustainable food choices for the future, we will be taking our first steps to protect our future as an industry. The first conference was so successful another is already being planned.

Middlebury College in Vermont has created an organic salad bar stocked almost entirely from vegetable gardens on campus, overseen by an on-staff dining services gardener using compost generated by the university's own food waste. The school's thirty- by forty-foot greenhouse provides common greens such as beet greens, kale, and arugula, as well as a variety of herbs and some more unusual greens such as kyona/mizuna, tatsoi, claytonia, and minutina. This is the kind of system that almost any school could implement with a little planning and passionate commitment.

In 1999 the first school dedicated solely to the marriage of sustainable agriculture and sustainable cuisine had its beginnings with a founder's dinner at Odessa Piper's restaurant, L'Etoile. The School of Organic Farming and Cooking at Taliesin is proposed to be housed on Frank Lloyd Wright's estate. Wright was known for his "organic architecture," and the 600-acre estate includes almost 200 acres historically utilized as farmland. The school will be a training ground for chefs and farmers schooled in the theory and practice of organic agriculture, foraging, and sustainable culinary practices. The goals of the school will be to convert the present conventional agricultural practices of the estate to organic/sustainable ones and to create a culinary school environment that places growers and culinary students together to learn and apply the knowledge of growing, harvesting, and cooking seasonal ingredients. We must look to this type of alternative culinary institution as a way to teach the next generation of culinarians.

Of course, schools and colleges are not the only organizations that provide valuable culinary education. Other groups we should support include Chefs Collaborative 2000, the American Livestock Breeds Conservancy, Seed Savers Exchange, Native Seed Search, the Chef and the Child Foundation, Oldways, SOS, Mothers & Others for a Livable Planet, and the Fresh Network, as well as corporations who make it a priority to keep consumers informed. Stonyfield Farms, Ben & Jerry's, and the Organic Cow are a few national companies that come to mind, but do your own research into local producers. Support them, and tell your friends why.

GREEN TAXES

The idea of green taxes—essentially, a set of federally imposed taxes on pollutants, pesticides, herbicides, fossil fuels, and other environmentally damaging materials used by businesses or private citizens—has been around for many years. Proposed regulations stipulate that the proceeds would fund environmental research or support businesses that focus on environmental stewardship. Green taxation would work to level the playing field for small farmers: larger agribusiness farms tend to use more fossil fuels and more chemicals than small farms, and forcing them to pay more to operate will drive the cost of food up, decreasing the advantage they currently hold. Paul Hawken writes in *The Ecology of Commerce* that green taxes would allow the marketplace to "be restored to its oft-praised purpose in life, which is to sort out the winners and losers. The winner will be the farmer who best takes care of his or her soil, animals, and posterity, not corporate entities that are essentially mining and extracting fertility for short-term gains.... In the upside-down and inverted logic of the present economic system, we cannot imagine that there is a point where something is too cheap. America is proud that its citizens pay the lowest percentage of disposable income for its groceries, but as the man at the farmer's market always tells me, you get what you pay for."

The difficulty of calculating the ultimate health and environmental costs associated with the use of chemicals has made green taxes tricky to implement. Quite a bit of research is available on the indirect or hidden costs of pesticides, however, and it may indeed be feasible to calculate a tax for pesticide use. In fact, some studies suggest that pesticide usage costs us over $8 billion a year in health and environmental costs. Research has also been done on the cost of fossil fuels and other environmental damage, a type of degradation we must start to tax if we are to rebuild the health of the planet.

Hawkens continues: "Imagine, if you will, paying 20 percent more for your food than you do now. Then imagine that the 20 percent is essentially credited back through reductions in income tax. Now imagine more family farmers, healthier food with less or no toxins on or within it, more gainful and meaningful employment in rural areas, and greater access to a wider variety of fresh foods."[15] It is for these reasons that I strongly support some form of green taxes. We should all speak to our representatives, educate them if necessary, and support the long-term health of our food supply by paying a little more now.

CELEBRATING FLAVORFUL SUSTAINABLE CUISINE

Chefs influence our food choices, and it is their responsibility to share their passion for sustainable, regional, seasonal food with other chefs, young culinarians, and the public. Chef and author Joyce Goldstein believes that it is crucial to celebrate artisan food producers in the same way we have come to celebrate chefs: "We are all going to be eating the same crap all over the world if we do not take the time to celebrate, for example, artisan chees makers, and local people that are making olive oil. We need to make the farmer glamorous, the cheese maker glamorous, the olive oil maker glamorous. We need to get these people to look like they have the sexiest job in the world, and then there will be more appeal and more security that you can spend your time doing this, and have a market for your wares. The restaurants will lead the way!"[16]

Rick Bayless balances his desire to cook using local ingredients with the demands of a northern city restaurant: "I believe so strongly in a direct connection with people who are supplying us that we work with a fruit farmer in Florida that is part of our extended family here. We work with a couple of farmers in Arizona and in California that we bring things in from, and while they are not local the products are seasonal to the places where they are grown. They are not being forced in any way. We have decided that since we run a year-round business we would try to have people supplying us during a period when we cannot have all of our stuff locally grown, who we really feel are part of our family."[17]

Odessa Piper's restaurant exemplifies an ideal, particularly in her processing and storing of seasonal foods for later use. Working this way often results in a greater burden on the labor force at certain times of the year (as you saw in chapter 2), but it can lead to some wonderfully creative meals, as well as a greater understanding of various food products by the restaurant staff.

The most valuable lessons we can take from Odessa's winter menus are in her creativity and willingness to utilize unusual products to their fullest potential. Odessa's culinary creations are proof that even in the harshest climates it is possible to produce enticing, flavorful foods using regional ingredients.

The Making of L'Etoile's Cuisine
Odessa Piper

On a personal level, I can think of six key experiences that brought L'Etoile to the place where we could produce the menus that we do. I would start with the way my parents raised me; we had our own garden, canned, and had many happy family outings centered around foraging. Then, in what would have been my last year of high school, I continued my higher education on a Luddite commune where we produced most of our food and stored it over the winter using rustic methods. In 1970 I moved to Wisconsin and worked with a visionary businesswoman named JoAnna Guthrie. She was attempting to link a working organic farm with a restaurant in Madison, both of which she started for that purpose. Though the restaurant was extremely popular, neither farm nor restaurant had the benefit of surrounding infrastructure to succeed as originally envisioned. At that point we had not addressed the idea of year-round regional reliance. By the standards of the 1970s this was something a scruffy band of communards could attempt to pull off but still unthinkable for a fine dining restaurant in the snow belt.

When I opened L'Etoile in 1976 I was integrating these earlier experiences into my menu strategies. The words "Cooking for the Seasons" were in my logo. The Dane County Farmers' Market across the street from us in Madison was entering its fourth year, but there was still no local infrastructure to allow us to purchase locally produced dry staples, freshwater fish, meat, or poultry on the scale a restaurant needed, if at all. Like many of the farmers who were attempting to practice organic agriculture at the time, we had altruistic goals, little experience, few mentors, and no societal support. I remember these times as like having to build a boat while learning to sail it, all the while bailing constantly to keep from drowning.

By 1980 we were routinely relying on winter-keeping vegetables and many varieties of fruit that we processed or froze for winter use. Our menus were always seasonal, but in winter we were still dependent on California for green vegetables. We had local eggs and an erratic supply of free-range chicken (inadequate processing infrastructure) and experimented with buying a whole milk-fed veal calf from an Amish family. That entire transaction was done by postcard because the Amish do not use phones. Then in 1983 on a retreat at Chinook Learning Center on Whidbey Island off the coast of Seattle I had an epiphany. For two weeks in the middle of summer our group had eaten almost exclusively from the center's garden and surrounding region. I directly experienced how the foods we ate shared their power of place with us and offered healing and wholeness. This octave of nutrition had been split off from the practice of agriculture in our "any thing-any-time-anywhere" culture, but it could be returned by choosing to cook with native and adapted species and voluntarily staying within regional and climactic limitations. This is by no means an original insight—all the great cuisines of the world are founded on this tenet—but it deepened my conviction of what we could aim for at L'Etoile.

Ten years later we still had only an erratic supply of locally produced meats, but local vegetable farmers were experimenting with wintered-over crops held under unheated hoop houses and cellaring loads of squash and roots, and we bought everything they had. Locally grown leafy vegetables, roots, syrups, nuts, eggs, poultry, cheeses, and fresh and preserved fruits made up over 60 percent of our winter menu. This last year, with the exception of leafy greens, about 80 percent of our vegetables were locally grown. Our farmers continue to figure out low-tech, affordable ways to supply us with greens through the snow months, and I think this coming winter will be the breakthrough. The next key shift, around 1993, involved a wonderful group of people here at the Center for Integrated Agricultural Systems, UW-Madison. Steve Stevenson, Jack Kloppenburg, and John Hendrickson, among others, were identifying and working with the ideas of bio-regionalism, food sheds, and regional reliance. Their concepts and contexts helped me explain to others as well as myself this fuzzy revolution we were in. The center also took on some of the thorny issues of infrastructure and began providing essential support through interdisciplinary seminars, research, and feasibility studies.

Due in no small part to their efforts, almost all of our poultry and meats (lamb, bison, beef, pink veal) are now directly supplied by area farmers. Ironically, we have to get our humanely raised pork from the Midwest via a wholesaler in California. I must emphasize here that we have never been, nor will ever be, regionally exclusive. The completely regional menu that follows could be drawn from a typical winter menu. But in order to stay on the fine dining radar, there always will be some ingredients that we bring in from outside our region; wine is a good example, ocean fish, citrus, or delicate salad greens. The point is, as much as possible the ingredients we select for L'Etoile's menu are from the Midwest or come to us in our winter having been grown in native soils and harvested in their natural cycles. Shad roe during their run, yes! Meyer lemons, yes!

Most recently, I would acknowledge the insights and opportunities that my husband, Terry Theise, has contributed to this process. Terry selects and imports wine to this country from small family wineries in Europe. He likes to talk about how these wines can impart their spirit of place to us, and how this can help us locate ourselves in an increasingly placeless world. Through getting to know the vintners and their dedication to the land, I have come to understand that these artisans are also farmers and that farmers can also be artisans. These extraordinary people have taken us into their homes and cooked for us and revealed to us the whole context of love and fealty and art that goes into the making of their wines. They and most Europeans enjoy everyday access to artisanal food and wines, beautiful cities, and accessible countryside, all anchored in a sense of place and a cultural appreciation of limits. They have preserved the culture in agriculture and have great food to show for it. It is like going back to what could be our future.

A Sample February/March Five-Course Dinner Menu at L'Etoile

Puree of Five White Root Vegetables with Watercress
and Beauty Heart Radish Compound Butter
Jerusalem artichokes and parsnips freshly dug from specially prepared trenches are particularly crisp and sweet. Celeriac, potato, beauty heart radish, and horseradish root are from the root cellar at Harmony Valley Farm and brought in with other long-keeping vegetables such as onions, garlic, shallots, and beets over the course of the winter. Watercress is gathered at stream heads where moving water keeps the leaves green, even under the snow, all winter long, and by early March it is prolific and we offer whole salads of it to our guests.

Rainbow Trout with Beurre Noisette of Wild Butternuts, Ground-cherries,
and Wintered Leeks Served on Wild Rice and Seared Spinach
The trout is locally farmed with artesian spring water. Butter from the Kickapoo region is browned with the green tops of wintered-over leeks from Kingsfield Garden. The leeks are banked and kept out in the field; as the days lengthen, new growth starts and is protected by the outer sheaths of last year's growth. The new growth is very sweet and tender—we use the entire plant. Wild butternuts are gathered by the old-timers, who enjoyed them for their crisp starchy texture. Ground cherries are an indigenous prairie plant that has been cultivated. They are bright orange, a color that stimulates weary winter appetites, and not too sweet for inclusion in savory applications. Wild rice comes from upstate; the pilaf includes braised white sections of the leeks, and the spinach is kept in the field throughout the winter in plastic tunnels with no additional heating.

> ## The Making of L'Etoile's Cuisine (continued)
>
> *Salad of Miner's Lettuce and Johnny Jump-ups*
> The miner's lettuce (also called winter purslane) grows slowly in unheated greenhouses through December, stays dormant (and harvestable) through the coldest months, and starts growing again in February. By March it is abundant, and bears edible flowers. The vinaigrette uses local vegetable oil and vinegar made from Seyval grapes grown in the region.
>
> *Warm Apple Tart with Candied Rose Petals and Sweet Charlie*
> *and Chèvre Ice Cream*
> The apples, both russets and Ida reds, are kept through April. The first chèvre comes from Fantøme in Ridgeway after the firstborn kids have their fill; the cheese is particularly creamy and tangy and must be used quickly, as it must be not salted for this dessert application. The rest of the ice cream base is made with Kickapoo Organic Valley heavy cream and organic eggs from Sweet Earth Farm. The sugar is from Wisconsin cane beets, and the candied leaves of herbs and flowers are sugared and dried over the course of the summer.
>
> *Whole Brandwine Raspberry Truffles Dipped in Chocolate*
> Four varieties of bramble berries are selected for perfect ripeness, size, and beauty and individually frozen; come winter we dip them in dark chocolate. The chocolate is local, the coffee, not.

VOTE BECAUSE YOU CAN MAKE A DIFFERENCE

Alice Waters recently said, "Eating is a political act." Wes Jackson commented, "Eating is an agricultural act." I believe there is truth in both statements. When we shop for food, we cast a vote for the future of our planet. When we buy processed foods, we vote for agribusiness. Fresh local food racks one up for your neighborhood farmer; eating at McDonald's supports agribusiness, while dining at restaurants like Chez Panisse keeps your money local. In early 1999 our voting potentially made a difference with the proposed organic standards regulations. The Associated Press reported that even though the new rules have not yet been resubmitted, a number of issues are being resolved, such as prohibiting the label "organic" for any food that is irradiated, genetically engineered, or treated with antibiotics. Additionally, the rules will also allow regional organic certification offices to certify meat as organic, something that the national government has yet to do.

Since the safety of our food supply is our government's responsibility, it is our job to let our representatives know when proposed or existing rules and regulations are not to our satisfaction. It takes a little work to stay on top of the issues, but the end result is worth the extra effort. A number of issues presented here are in the news almost every day. Others are hidden in the back pages of the national press.

Each of us, every day, makes a difference—good or bad—simply through the act of eating. Looking at the big picture can seem overwhelming, and that is why I hope that the information in this book, the ideas, and the stories have provided you with the knowledge you will need to begin to make informed choices about the food you eat. I urge you to take stock of your pantry and start making changes as soon as possible so that we might have a food supply that is sustainable for generations to come. As Kahlil Gibran puts it, "A little knowledge that *acts* is worth infinitely more than much knowledge that is idle."

A Checklist for Making Your Vote Count

- Always buy local whenever possible, from as close a source as possible.
- Support farmers' markets and farmstands.
- Join a CSA.
- Next time you are in your supermarket, ask to talk to the produce manager. Tell the manager of your concern about pesticides, and say you would prefer to buy local regional produce and certified organic food if possible.
- Try regional products new to you.
- Plant a garden and help set up a school garden.
- Ask your grocers what type of farms the meat and poultry they sell is raised on. If they do not know, ask them to find out.
- Support grocers and butchers who get their supplies from farmers who do not use factory-farming techniques.
- Eat seasonally; do not buy out-of-season produce.
- Read labels, to find out what is in the food you are eating.
- Educate yourself, understand the issues, and let your legislators know how you feel and what rules and regulations are important to you.

Appendix A
Organizations and Resources

AGRICULTURAL RESEARCH SERVICE

ARS is the in-house research arm of the U.S. Department of Agriculture. The mission of the ARS is to develop and transfer solutions to agriculture problems of high national priority and provide information access and dissemination to ensure high-quality, safe food and other agricultural products, assess the nutritional needs of Americans, sustain a competitive agricultural economy, enhance the natural resource base and the environment, and provide economic opportunities for rural citizens, communities, and society as a whole.

website: *www.ars.usda.gov.afm/mr.html*

AMERICAN FARMLAND TRUST

The American Farmland Trust is a private, nonprofit organization founded in 1980 to protect our nation's farmland. AFT works to stop the loss of productive farmland and to promote farming practices that lead to a healthy environment. Its action-oriented programs include public education, technical assistance in policy development, and direct farmland protection projects.

American Farmland Trust
1200 18th Street, NW, Suite 800
Washington, DC 20036
202–331–7300
fax: 202–659–8339
e-mail: *info@farmland.org*
website: *www.farmland.org*

AMERICAN LIVESTOCK BREEDS CONSERVANCY

Founded in 1977, ALBC is the only breed conservation organization in the United States. This nonprofit organization promotes and conserves rare breeds of livestock (horses, cattle, sheep, goats, pigs, and poultry). These animals, which were selected and developed by generations of farmers during an era that envisioned a sustainable future, embody the historical preeminence of North American agriculture and represent a genetic diversity necessary to meet the agricultural needs of the future. The ALBC conducts direct conservation projects, such as a rare-breeds semen bank and population rescues, and provides educational programs, assistance to breed associations, and information on rare breeds. The 3,500 members of ALBC share a commitment that rare breeds will survive.

American Livestock Breeds Conservancy
P.O. Box 477
Pittsboro, NC 27312
919–542–5704

CENTER FOR DEMOCRACY AND TECHNOLOGY

CDT works to promote democratic values and constitutional liberties in the digital age. With expertise in law, technology, and policy, CDT seeks practical solutions to enhance free expression and privacy in global communications technologies. CDT is dedicated to building consensus among all parties interested in the future of the Internet and other new communications media.

Center for Democracy and Technology
1634 I Street, NW, Suite 1100
Washington, DC 20006
202–637–9800
fax: 202–637–0968
website: *www.cdt.org*

CENTERS FOR DISEASE CONTROL AND PREVENTION

CDC's mission is to promote health and quality of life by preventing and controlling disease, injury and disability. CDC pledges to be a diligent steward of the funds entrusted to it, to provide an environment for intellectual and personal growth and integrity, to base all public health decisions on the highest-quality scientific data, openly and objectively derived, to place the benefits to society above the benefits to the institution, and to treat all persons with integrity, honesty, and respect.

Centers for Disease Control and Prevention
1600 Clifton Road, NE
Atlanta, GA 30333
800–311–3435 (public inquiries—toll free)
fax: 404–639–6290 (public inquiries)
e-mail: *infoply@cdc.gov*
website: *www.cdc.gov*

CENTER FOR URBAN AGRICULTURE AT FAIRVIEW GARDENS

The Center for Urban Agriculture at Fairview Gardens is situated on a hundred-year-old working organic farm in the heart of suburban Goleta, California. The twelve acres of farmland is owned by the nonprofit center, which has placed it under an agricultural conservation easement. This preserves Fairview Gardens as a sustainable organic farm in perpetuity and allows it to continue its vital educational programs. The center's mission is to preserve the farm's agricultural heritage, provide fresh, chemical-free fruits and vegetables to the local community, demonstrate the economic viability of sustainable agricultural methods for small farm operations, research and interpret the connections between food, land stewardship, and community well-being, and nurture the human spirit through educational programs and public activities at the farm.

Center for Urban Agriculture at Fairview Gardens
598 N. Fairview Avenue
Goleta, CA 93117
805–967–7369
fax: 805–967–0188
e-mail: *fairviewg@aol.com*

CHEFS COLLABORATIVE 2000

CC2000 promotes sustainable cuisine by teaching children, supporting local farmers, educating one another, and inspiring the public to choose good, clean food. Founded in 1993 by Oldways Preservation and Exchange Trust with a group of leading chefs, the Collaborative is a national nonprofit membership organization of chefs who are dedicated to the ethic of sustainable cuisine. Programs currently in place are farm-to-restaurant projects and the Adopt-a-School program, an in-school curriculum that teaches children about food, history, culture, and the environment.

Chefs Collaborative 2000
282 Moody Street
Waltham, MA 02154
781–736–0635
fax: 781–642–0307
e-mail: cc2000@chefnet.com
website: *www.chefnet.com/cc2000*

CONGRESSIONAL HUNGER CENTER

Democratic and Republican members of Congress formed the Congressional Hunger Center in 1993 after the Select Committee on Hunger was eliminated. The program is designed to train emerging leaders in the fight against hunger. Young adults are working across America on Native American reservations, in urban food banks, and at homeless shelters. Junior high school students are involved in finding solutions to hunger, including planting community gardens. Through the International Crisis Response Program, the center is working with existing leaders on solutions to the famine crisis in North Korea and refugee crises in Rwanda and Bosnia.

Congressional Hunger Center
229$^1/_2$ Pennsylvania Avenue, SE
Washington, DC 20003
202–547–7022
fax: 202–547–7575
e-mail: *nohunger@aol.com*
website: *www.hungercenter.org*

EARTH COMMUNICATIONS OFFICE

The Earth Communications Office, the environmental voice of the entertainment industry, is a nonprofit, nonpartisan organization that uses the power of communication to improve the global environment. It was formed ten years ago to serve as a bridge between the environmental community and the entertainment industry. ECO has created award-winning communication campaigns using movie theater, television, and radio public service announcements, that have been seen by over 1 billion people worldwide. ECO's goal is to educate and inspire people around the world to take action to help protect the planet.

Earth Communications Office
12021 Wilshire Blvd., #557
Los Angeles, CA 90025
310–656–0577
fax: 310–656–1657
e-mail: *ecoffice@earthlink.net*
website: *www.OneEarth.org*

EASTERN NATIVE SEED CONSERVANCY

Since 1992, the Conservancy has dedicated itself to the biocultural conservation of economic plant resources, especially food and medicinal plants, adapted or endemic to the northeastern United States, with special consideration to those plants of Native American origination or usage. The basis of the Conservancy's conservation agenda are is the CRESS Heirloom Seed Conservation Project (Conservation and Regional Exchange by Seed Savers) and the Native Seeds Project, a grassroots-oriented solution for the conservation of these resources in a sustainable manner, through seed collection, propagation, seed banking, and redistribution.

Eastern Native Seed Conservancy
222 Main Street, Box 451
Great Barrington, MA 01230
413–229–8316
website: *http://gemini.berkshire.net*

ECOLOGY ACTION CENTRE

The Ecology Action Centre has been an active advocate, protecting the environment, since 1972. The Centre's earliest projects included recycling, composting, and energy conservation, and these are now widely recognized environmental issues. Currently the Centre is focusing on marine, wilderness, and transportation issues and school ground naturalization, as well as serving as intervenors on the proposed Sable Offshore Energy Project.

Ecology Action Centre
1568 Argyle Street, Suite 31
Halifax, Nova Scotia
Canada B3J 2B3
902–429–2202
fax: Please call ahead
e-mail: *www.chebucto.ns.ca/Environment/EAC/EAC-Home.html*

EDIBLE SCHOOLYARD

Through the hard work of almost 900 students and their teachers, the Edible Schoolyard garden has been transformed from a cluster of beds to a beautiful and productive garden. The vision of the program is providing a garden-to-kitchen-to-table experience for students. Alice Waters provides insight and guidance throughout the year, supporting the program through her collaboration with the Martin Luther King Middle School and the Edible Schoolyard. The mission of the Edible Schoolyard at King Middle School is to create and sustain an organic garden and landscape that is wholly integrated into the school's curriculum and lunch program and involves students in all aspects of farming the garden, along with preparing, serving, and eating the food, as a means of awakening their senses and encouraging awareness and appreciation of the transformative values of nourishment, community, and stewardship of the land.

Edible Schoolyard
Martin Luther King Jr. Middle School
1781 Rose Street, Berkeley, CA 94703
510–558–1335
fax: 510–558–1334
e-mail: *edible@lanminds.com*

ENVIRONMENTAL DEFENSE FUND

The Environmental Defense Fund was founded in 1967 by ten volunteer conservationists on Long Island concerned with the decline of osprey nesting sites on local beaches. Today EDF is an effective environmental force backed by more than 300,000 members across America. The combination of science, economics, law, and public activism that EDF brings to bear on all environmental problems has led to numerous landmark victories over the years. Concern for the planet's health and a shared belief that economically viable solutions do exist for even the most perplexing environmental problems form the common bond of EDF members.

Environmental Defense Fund
1875 Connecticut Avenue, NW
Washington, DC 20009
800–684–3322
202–387–3525
e-mail: *members@edf.org*
website: *www.edf.org*

ENVIRONMENTAL PROTECTION AGENCY

The Environmental Protection Agency, founded in 1970, implements the federal laws designed to promote public health by protecting our nation's air, water, and soil from harmful pollution. EPA endeavors to accomplish its mission systematically by proper integration of a variety of research, monitoring, standard-setting, and enforcement activities. As a complement to these, EPA coordinates and supports research and antipollution activities of state and local governments, private and public groups, individuals, and educational institutions. EPA also monitors the operations of other federal agencies with respect to their impact on the environment.

Environmental Protection Agency
401 M Street, SW
Washington, D.C. 20460–0003
202–260–2090
website: *www.epa.gov*

FOOD AND AGRICULTURE ORGANIZATION OF THE UNITED NATIONS

Founded in 1945, the FAO, an autonomous agency within the UN system, is the leading international body for food and agriculture. It has a membership of 175 countries—plus the European Community—which have pledged to raise levels of nutrition and standards of living of their peoples; to improve the production and distribution of all food and agricultural products; and to improve the condition of rural people. The FAO also provides advice and assistance to its member countries on national commodity and trade policy problems and oversees the Global Information and Early Warning System on Food and Agriculture, which monitors global supply and demand for basic foodstuffs and fertilizers.

FAO Headquarters
Viale delle Terme di Caracalla
00100 Rome, Italy
[39] (06) 5–225–5225
telex: 610181 FAO I
e-mail: *telex-room@fa*
website: *www.fao.org*

THE FOOD AND DRUG ADMINISTRATION (FDA)

The FDA, one of our nation's oldest consumer protection agencies, is an agency within the Public Health Service, which in turn is a part of the Department of Health and Human Services (HHS). FDA ensures that the food we eat is safe and wholesome, that the cosmetics we use won't harm us, and that medicines, medical devices, and radiation-emitting consumer products such as microwave ovens are safe and effective. FDA also oversees feed and drugs for pets and farm animals. Authorized by Congress to enforce the Federal Food, Drug and Cosmetic Act (FFDCA) and several related public health laws, the agency monitors the manufacture, import, transport, storage, and sale of about $1 trillion worth of products each year.

The Food and Drug Administration
5600 Fishers Lane
Rockville, MD 20857
1–888–463–6332
website: *www.fda.gov*

GREEN GULCH FARM ZEN CENTER

Green Gulch Farm Zen Center, also known as Green Dragon Temple (Soryu-ji), is a Buddhist practice center in the Japanese Soto Zen tradition offering training in Zen meditation and ordinary work. In addition to the temple program of meditation and study, it includes an organic farm and garden—Green Gulch Farm—as well as a guest and conference center. The center offers organic gardening classes, tea classes, meditation retreats, and workshops, many designed for families. The Zen Center's Mountain Gate Study Center offers a variety of classes in Buddhist philosophy and practice.

Green Gulch Farm Zen Center
1601 Shoreline Highway
Sausalito, CA 94965
415–383–3134
fax: 415–383–3128
website: *http://bodhi.zendo.com*

GREENPEACE

Greenpeace proclaims that its purpose is to create a green and peaceful world. It believes that determined individuals can alter the actions and purposes of even the most powerful by "bearing witness," that is, by drawing attention to an abuse of the environment through their unwavering presence at the scene, whatever the risk.

Greenpeace National Office
1436 U Street, NW
Washington, DC 20009
800–326–0959
fax: 202–462–4507
e-mail: *greenpeace.usa@wdc.greenpeace.org*
website: *www.greenpeace.org*

HEIFER PROJECT INTERNATIONAL

With more than fifty years experience, Heifer Project has proved its approach to helping people obtain a sustainable source of food and income, by giving families a source of food rather than short-term relief. Heifer Project works with grassroots community groups who determine their own needs, and train and prepare for their animals. HPI trains farmers to manage grazing, plant trees and crops, and use natural fertilizer to improve the environment. The project also funds more than eighty Women in Livestock Development projects, which provide women with food- and income-producing animals as well as training in leadership, community development, and environmentally sound farming.

Heifer Project International—World Headquarters
P.O. Box 808
Little Rock, AR 72203
800–422–0474
fax: 501–376–8906
e-mail: *donorservices@heifer.org*
website: *www.heifer.org*

INSTITUTE FOR AGRICULTURE AND TRADE POLICY

Established in 1986, the Institute for Agriculture and Trade Policy is a non-profit research and education organization that creates environmentally and economically sustainable communities and regions through sound agriculture and trade policy. Through monitoring, analysis and research, education, training and technical assistance, and coalition building and international networking, the IATP assists public interest organizations in effectively influencing both domestic and international policymaking.

Institute for Agriculture and Trade Policy
2105 First Avenue South
Minneapolis, MN 55404
612–870–0453
fax: 612–870–4846
website: *www.iatp.org*

INTERVALE FOUNDATION

The Intervale in Burlington, Vermont, is 700 acres of floodplain alongside the Winooski River. Since 1988, the Intervale Foundation has worked locally to restore the Burlington Intervale as a vital agricultural, social, economic, and natural resource by making land available for new farmers, increasing their access to needed resources, and supporting local market growth. The foundation is an advocate for organic agriculture, innovative environmental solutions, and support for the local economy so everyone may have a healthy life.

Intervale Foundation
128 Intervale Road
Burlington, VT 05401
802–660–3508
fax: 802–660–3501
e-mail: Green City Farm—*farm@gardeners.com*
Intervale Compost Project—*compost@together.net*

JOHNNY'S SELECTED SEEDS
Johnny's Selected Seeds, a seed producer and merchant in Albion, Maine, was established in 1973 by founder and chairman Rob Johnston, Jr. Johnny's sells both retail and wholesale vegetable seeds, medicinal and culinary herb seeds, flower seeds, and high-quality tools, equipment, and gardening accessories. Johnny's operates its own certified organic trial and research farm in Albion for production and to grow, breed, and evaluate over 4,000 new and experimental varieties each year

Johnny's Selected Seeds
Foss Hill Road
Albion, ME 04910
207–437–4357
fax: 800–437–4290
e-mail: *johnnys@johnnyseeds.com*
website: *www.johnnyseeds.com*

LAND STEWARDSHIP PROJECT
The Land Stewardship Project, founded in 1982, is a private, nonprofit membership organization devoted to fostering an ethic of stewardship toward farmland. It is working to develop and promote sustainable communities and a system of agriculture that is environmentally sound, economically viable, family-farm-based, and socially just. LSP believes that all people—farmers and non-farmers alike—have a fundamental responsibility to care for the land that sustains us.

Land Stewardship Project
2200 4th Street
White Bear Lake, MN 55110
612–653–0618
website: *www.misa.umn.edu/lsphp*

MONSANTO
A life sciences company, Monsanto is committed to finding solutions to the growing global needs for food and health by sharing common forms of science and technology among agriculture, nutrition, and health. Monsanto's

30,000 employees worldwide seek to make and market high-value agricultural products, pharmaceuticals, and food ingredients in a manner that achieves environmental sustainability.

Monsanto
800 North Lindbergh Boulevard
St. Louis, MO 63167
314–694–1000
website: *www.monsanto.com*

MOTHERS & OTHERS FOR A LIVABLE PLANET

Mothers & Others, a nonprofit consumer education and advocacy organization, seeks to effect lasting protection of children's health and the environment. Driven by the belief that environmental change begins at home, Mothers & Others is working to bring about a shift in overall consumption patterns by focusing the collective marketplace power of women, among others, on choices that are healthy, safe, and environmentally sound for their families and communities. Mothers & Others was established in 1989 by Meryl Streep and her Connecticut neighbors out of concern that children were being exposed to unsafe pesticide levels in the food supply.

Mothers & Others for a Livable Planet
40 West 20th Street
New York, NY 10011
212–242–0010, ext. 307
fax: 212–242–0545
e-mail: *mothers@mothers.org*
website: *www.mothers.org*

NATIONAL CAMPAIGN FOR SUSTAINABLE AGRICULTURE

The National Campaign for Sustainable Agriculture is a network of over 500 diverse groups whose mission is to shape national policies to foster a sustainable food and agricultural system—one that is economically viable, environmentally sound, socially just, and humane. The goals of the campaign are to foster understanding, communication, and leadership development, to engage in policy analysis and development, and to work together to advance policy objectives.

National Campaign for Sustainable Agriculture
P.O. Box 396
Pine Bush, NY 12566
914–744–8448
fax: 914–744–8477
e-mail: *campaign@magiccarpet.com*
website: *www.SustainableAgriculture.net*

NATIONAL GARDENING ASSOCIATION

Founded in 1972, NGA is the largest nonprofit gardening organization in the country. It helps gardeners and nongardeners alike through its high-quality publication *National Gardening* magazine, its innovative science education programs, and its garden-related research. NGA has reached out with gardening to revitalize communities, feed the hungry, teach respect for the environment, and provide educators with innovative and effective teaching methods and materials through the *Growing Ideas Catalog*.

National Gardening Association
180 Flynn Avenue
Burlington, VT 05401
800–538–7476
802–863–1308
fax: 802–863–5962
e-mail: *eddept@garden.org*
website: *www.garden.org*

NATIONAL INSTITUTES OF HEALTH

NIH began as a one-room laboratory of hygiene in 1887; today it is one of the world's foremost biomedical research centers, and the federal focal point for biomedical research in the United States. NIH's mission is to uncover new knowledge that will lead to better health for everyone. It works toward that mission by conducting research in its own laboratories; supporting the research of nonfederal scientists in universities, medical schools, hospitals, and research institutions throughout the country and abroad; helping in the training of research investigators; and fostering communication of biomedical information.

National Institutes of Health
9000 Rockville Pike
Bethesda, MD 20892
301–496–1776
website: *www.nih.gov*

THE NATIONAL MARINE FISHERIES SERVICE

The National Marine Fisheries Service, or "NOAA Fisheries," is a part of the National Oceanic and Atmospheric Administration (NOAA). NMFS administers NOAA's programs, which support the domestic and international conservation and management of living marine resources. NMFS provides services and products, to support domestic and international fisheries management operations, fisheries development, trade and industry assistance activities, enforcement, protected species and habitat conservation operations, and the scientific and technical aspects of NOAA's marine fisheries program.

The National Marine Fisheries Service
National Oceanic and Atmospheric Administration
1315 East-West, Hwy SSMC 3
Silver Spring, MD 20910–3226
301–713–2370
fax: 301–713–1452
website: *www.nmfs.gov*

NATIONAL SEED STORAGE LABORATORY

The National Seed Storage Laboratory has been in operation since 1958. It is part of the National Plant Germplasm System (NPGS), which collects, documents, preserves, evaluates, enhances, and distributes plant genetic resources for continued improvement in the quality and production of economic crops important to U.S. and world agriculture. The mission of the NSSL is to preserve the base collection of the NPGS and to conduct research to develop new technologies for preservation of seed and other propagules of plant genetic resources.

National Seed Storage Laboratory
111 South Mason Street
Fort Collins, CO 80521–4500
970–495–3200
fax: 970–221–1427
website: *www.ars-grin.gov/nssl/nsslmain.html*

NATURAL RESOURCES DEFENSE COUNCIL

The Natural Resources Defense Council works to establish sustainability and good stewardship of the earth as central ethical imperatives of human society. NRDC affirms the integral place of human beings in the environment, and strives to protect nature in ways that advance the long-term welfare of present and future generations. NRDC is a conservation force on many fronts, addressing problems such as are depleting ocean fish populations: overfishing, industrial pollution and runoff, and bycatch.

Natural Resources Defense Council
1200 New York Avenue, NW, Suite 400
Washington, DC 20005
202–289–6868
e-mail: *nrdcinfo@nrdc.org*
website: *www.nrdc.org*

NATIVE SEEDS/SEARCH

Founded in 1983, Native Seeds/SEARCH is a nonprofit organization preserving Native American crop seeds. NS/S works to conserve the traditional crops, seeds, and farming methods that have sustained native peoples throughout the southwestern United States and northern Mexico. A major regional seed bank and a leader in the heirloom seed movement, NS/S works to protect biodiversity and to celebrate cultural diversity, through research, training, and community education.

Native Seeds/SEARCH
2509 N. Campbell Avenue #325
Tucson, AZ 85719
Fax: 602–327–5821
website: *www.desert.net/seeds*

NORTHEAST ORGANIC FARMING ASSOCIATION

NOFA is a nonprofit association of consumers, gardeners, and diversified farmers who share a vision of local, organic agriculture, working together through education and membership participation to strengthen agriculture in Vermont. As a NOFA member, you may apply for organic certification through Vermont Organic Farmers (VOF), the certification committee of NOFA. VOF certifies livestock, vegetable, fruit, and dairy producers.

NOFA of Vermont
15 Barre Street
Montpelier, VT 05602
802–229–4940
website: *www.nofavt.org*

OLDWAYS PRESERVATION AND EXCHANGE TRUST

Oldways, a nonprofit organization founded in 1990, promotes healthy eating based on the traditional cuisines of cultures all over the world using food grown in environmentally sustainable ways. Its mission is to educate people about healthy foods, encouraging them to change the way they eat and to move toward greater health and a healthier planet, through healthy eating guides, conferences, publications, and educational programs. Oldways is also a partner of Chefs Collaborative 2000, a national network of more than 1,500 leading chefs.

Oldways Preservation and Exchange Trust
25 First Street
Cambridge, MA 02141
617–621–3000
fax: 617–621–1230
e-mail: *oldways@tiac.net*
website: *www.oldwayspt.org*

THE ORGANIC TRADE ASSOCIATION

The OTA is a national association representing organic industry in Canada and the United States. Members include growers, shippers, processors, certifiers, farmer associations, brokers, consultants, distributors, and retailers.

Established in 1985 as the Organic Food Production Association of North America, the OTA works to promote organic products in the marketplace, protect the integrity of organic standards, increase the sales and sustainability of the industry, protect the environment and sustain a balanced ecosystem, and provide a strong and unified voice on legislative, regulatory, and policy issues.

Organic Trade Association
50 Miles Street
P.O. Box 1078
Greenfield, MA 01302
413–774–7511
fax: 413–774–6432
e-mail: *ota@igc.apc.org*
website: *www.ota.com*

ORGANIC FARMING RESEARCH FOUNDATION

The Organic Farming Research Foundation, a nonprofit foundation directed by certified organic farmers, was founded in 1990, with a board of directors composed of organic farmers, researchers, and activists from around the nation. The OFRF is the only organization in the nation supporting organic farming research and education through grants and advocacy. OFRF's mission is to sponsor research related to organic farming practices, to disseminate research results to organic farmers and to growers interested in adopting organic production systems, and to educate the public and decision makers about organic farming issues.

Organic Farming Research Foundation
P.O. Box 440
Santa Cruz, CA 95061
831–426–6606
fax: 831–426–6670
e-mail: *research@ofrf.org*
website: *www.ofrf.org*

PEAT INSTITUTE

The PEAT Institute provides strategic planning, organizational development, and issues management. Much of its work involves marrying corporate clients with the principles of environmental stewardship over marine and land-based resources. The institute also promotes partnerships among academic, corporate, and governmental entities to provide real-world environmental solutions.

PEAT Institute
611 Pennsylvania Avenue, SE, Suite 372
Washington, DC 20003–4303
202–544–9748
fax: 202–544–9749
e-mail: *PEATIns@Compuserve.com*

PUBLIC MARKET PARTNERS

Public Market Partners is a tax-exempt, not-for-profit organization that works in partnership with communities to plan, develop, and manage public markets. Since 1991, Public Market Partners has helped create new markets and assist established markets throughout the United States. Public Market Partners divides its activities into three program areas: technical assistance, development and management, and education and research. Projects have ranged from farmers' markets and market halls to assistance for the Watershed Agricultural Council, marketing and development of the Catskills Family Farm project, establishing open-air markets located in low-income communities in New York City, and organizing international conferences on public markets.

Public Market Partners
5454 Palisade Avenue
Bronx, NY 10471
212–524–7133

RURAL ADVANCEMENT FOUNDATION INTERNATIONAL

RAFI is an international nongovernmental organization dedicated to the conservation and sustainable improvement of agricultural biodiversity, and to the socially responsible development of technologies useful to rural societies. RAFI is concerned about the loss of genetic diversity—especially in agriculture—and about the impact of intellectual property rights on agriculture and world food security.

RAFI—USA
P.O. Box 640
Pittsboro, NC 27312
919–542–1396
fax: 919–542–0069
e-mail: *www@rafti.org*
website: *www.rafi.ca*

SEAWEB

SeaWeb is a project designed to raise awareness of the world ocean and the life within it. SeaWeb's goal is to educate and raise awareness of and interest in the ocean, making its continued well-being a priority for citizens of the United States and the world. Through public polling, SeaWeb monitors changes in awareness of and care for the ocean; it also sponsors or produces educational programs and announcements on radio, television, and film. SeaWeb is creating an independent ocean information center that reaches out to the media, to government officials, and to the interested public.

SeaWeb
1731 Connecticut Avenue, NW, 4th Floor
Washington, DC 20009
888–4–SEAWEB
e-mail: *tburley@seaweb.org*
website: *www.seaweb.org*

SECOND HARVEST

Second Harvest, established in 1979, is a network of food banks that distributes millions of pounds of donated food and grocery products to the hungry through food pantries, soup kitchens, and homeless shelters.

Second Harvest
116 S. Michigan Avenue, Suite 4
Chicago, IL 60603
312–263–2305
800–532–3663
fax: 312–236–5626

SEED SAVERS EXCHANGE

Founded in 1975 by Kent and Diane Whealy, Seed Savers Exchange is a nonprofit, grassroots organization with a network of 8,000 gardeners, orchardists, and plant collectors who are maintaining heirloom varieties of vegetables, fruits, grains, flowers, and herbs, which they distribute through SSE's annual publications. Seed Savers Exchange provides the model for organizations and genetic preservation projects in more than thirty countries.

Seed Savers Exchange
3076 North Winn Road
Decorah, IA 52101
319–382–5990
fax: 319–382–5872

SHARE OUR STRENGTH

SOS works to alleviate hunger and poverty in the United States and around the world. Since its founding in 1984, SOS has distributed more than $50 million in over 1,000 antihunger and antipoverty efforts worldwide. Operation Frontline, SOS's nutrition education and food budgeting program, connects chefs with people who are at risk from poor nutrition and hunger. Kids Up Front, a partnership with Kraft Foods, is the children's component of Operation Frontline; it helps children at risk of hunger to make better food choices and improve their diets over the long term.

Share Our Strength
1511 K Street, NW, Suite 940
Washington, DC 20005
800–969–4767
202–393–2925
fax: 202–347–5868
website: *www.strength.org*

SUSTAINABLE AGRICULTURE NETWORK

The Sustainable Agriculture Network is a cooperative effort of university, government, farm, business, and nonprofit organizations dedicated to the exchange of scientific and practical information on sustainable agricultural systems. The network supports the exchange of information with a variety of printed and electronic communication tools.

Sustainable Agriculture Network
National Agricultural Library
10301 Baltimore Blvd., Room 304
Beltsville, MD 20705
310–504–6425
fax: 301–504–6409
e-mail: *san@nalusda.gov*

SUSTAINABLE AGRICULTURE RESEARCH AND EDUCATION

SARE—the USDA's Sustainable Agriculture Research and Education program—works to increase knowledge about agricultural practices that are economically viable, environmentally sound, and socially acceptable. To advance such knowledge nationwide, SARE administers a competitive grants program first funded by Congress in 1988. SARE began offering a small grants program for farmers and ranchers to run their own onsite research experiments in 1992.

SARE National Office
USDA
Room 3868 South Bldg., Ag Box 2223
Washington, DC 20250–2223
202–720–5203
website: *www.sare.org*

THE CHEF AND THE CHILD FOUNDATION

In 1988 the American Culinary Federation designed a program addressing the nutritional and dietary needs of children, forming a nonprofit corporation, the Chef and the Child Foundation, Inc. The purpose of the foundation is to foster, promote, encourage, and stimulate an awareness of proper nutrition in preschool and elementary school children through education, community involvement, and fundraising.

The Chef and the Child Foundation
The American Culinary Federation
10 San Bartola Road
St. Augustine, FL 32086
904–824–4468

THE COOK'S GARDEN

The Cook's Garden began as an organic market garden supplying unusual and flavorful vegetables from around the world to local chefs and cooks. Many of its customers wanted to be able to grow some of these specialties in their own gardens, so in 1983 it began to sell seed, first at its farm stand, then from a catalog, and finally on the Internet. The gardens are located in Burlington, Vermont.

The Cook's Garden
P.O. Box 5010
Hodges, SC 29653–5010
800–457–9703
fax: 800–457–9705
e-mail: *gardens@cooksgarden.com*
website: *www.cooksgarden.com*

THE GARDEN PROJECT

The Garden Project, a postrelease program founded in 1983 in San Francisco, seeks to impact the local environment while offering prisoners and former prisoners an alternative to the cycle of crime and poverty.

The Garden Project
Pier Twenty-Eight
The Embarcadero
San Francisco, CA 94105
415–243–8558
fax: 415–243–8221

THE INSTITUTE FOR FOOD AND DEVELOPMENT POLICY (FOOD FIRST)

The Institute for Food and Development Policy, better known as Food First, is a member-supported, nonprofit "peoples" think tank and education-for-action center. The institute's work highlights root causes and value-based solutions to hunger and poverty around the world, with a commitment to establishing food as a fundamental human right. Founded in 1975 by Frances Moore Lappé and Joseph Collins, Food First provides leadership for reforming the global food system from the bottom up, offering an antidote to the myths and obfuscations that make change seem difficult to achieve.

The Institute for Food and Development Policy (Food First)
398 60th Street
Oakland, CA 94618
510–654–4400
fax: 510–654–4551
e-mail: *foodfirst@foodfirst.org*
website: *www.foodfirst.org*

U.S. DEPARTMENT OF AGRICULTURE
The mission of the USDA is to enhance the quality of life for the American people by supporting production of agriculture, thus ensuring a safe, affordable, nutritious, and accessible food supply; caring for agricultural, forest, and range lands; supporting sound development of rural communities; providing economic opportunities for farm and rural residents; expanding global markets for agricultural and forest products and services; and working to reduce hunger in America and throughout the world.

U.S. Department of Agriculture
14th and Independence Avenue, SW
Washington, D.C. 20250
202–720–2791
website: *www.usda.gov*

VERMONT LAND TRUST
Vermont Land Trust had it beginnings in 1976, and today protects over 147,000 acres of conserved land, 200 operating farms, two state parks, town recreation areas, a coalition of conservationists, and housing and public interests. The organizations intends to hold and manage the land in ways that will further the environmental, economic, and community goals of the region.

Vermont Land Trust
8 Bailey Avenue
Montpelier, VT 05602
802–223–5234
fax: 802–223–4223
website: *www.vlt.org*

WALNUT ACRES ORGANIC FARMS
Walnut Acres, founded as a small family operation by Paul and Betty Keene in Penns Creek, Pennsylvania, in 1946, is one of the oldest and most respected organic growers and processors in the United States. Walnut Acres pioneered a mail-order distribution system for its organic foods in a time when health food stores were extremely scarce; it has played a major role in the growth and acceptance of the organic farming movement.

Walnut Acres Organic Farms
Penns Creek, PA 17862
800–433–3998
fax: 717–837–1146
website: *walnutacres.com*

WORLD SUSTAINABLE AGRICULTURE ASSOCIATION

The World Sustainable Agriculture Association, a nonprofit corporation established in 1991, is organized as a federation of autonomous and self-supporting
branches in seven countries—the United States, India, Thailand, Taiwan, Japan,
Australia, and China. Branch activities vary widely from one country to
another, but include such services as research and demonstration farms, educational programs, assistance to farmers in transition to sustainable farming
systems, and information transfer by publications and electronic network.

World Sustainable Agriculture Association
8554 Melrose Avenue
West Hollywood, CA 90069
310–657–7202
fax: 310–657–3884
e-mail: *wsaala@igc.apc.org*
website: *www.igc.apc.org*

WORLDWATCH INSTITUTE

The Worldwatch Institute is a nonprofit public policy research organization
dedicated to informing policy makers and the public about emerging global
problems and trends and the complex links between the world economy and
its environmental support systems. The institute is dedicated to fostering the
evolution of an environmentally sustainable society in which human needs
are met in ways that do not threaten the health of the natural environment
or the prospects of future generations.

Worldwatch Institute
1776 Massachusetts Avenue, NW
Washington, DC 20036–1904
202–452–1999
fax: 202–296–7365
e-mail: *worldwatch@worldwatch.org*
website: *www.worldwatch.org*

Appendix B
Websites

Company/Association/Organization	Website Address
Agricultural Research Service	www.ars.usda.gov
Alternative Farming Systems Information Center	www.nal.usda.gov
American Cancer Society	www.cancer.org
American Dietetic Association	www.eatright.org
American Farm Bureau	www.fb.com
American Farmland Trust	www.farmland.org
American Genetic Resources Alliance	www.amgra.org
American Seed Trade Association	www.amseed.com
Appropriate Technology Transfer for Rural Areas	www.attra.org
Ben & Jerry's	www.benjerry.com
Beyond Pesticides	www.ncamp.org
Black River Produce	www.blackriverproduce.com
California Certified Organic Farmers	www.ccof.org
Center for Agroecology and Sustainable Food Systems	http://zzyx.ucsc.edu/casfs
Center for Democracy and Technology	www.cdt.org
Centers for Disease Control	www.cdc.gov
Center for Nutrition Policy and Promotion	www.usda.gov/cnpp
Center for Responsive Politics	www.crp.org
Chefs Collaborative 2000	www.chefnet.com/cc2000
Chez Panisse Café and Restaurant	www.chezpanisse.com
Coleman Natural Products, Inc.	www.colemannatural.com
Congressional Hunger Center	www.hungercenter.org
Cornell University	www.research.cornell.edu
Corporate Agribusiness Research Project	http://home.earthlink.net

Diabetes Action Research and   www.daref.org/education
  Education Foundation
Eastern Native Seed Conservancy   http://gemini.berkshire.net
Ecology Action   www.crest.org/sustainable/
                                            ecology_action.html

Environlink—   www.envirolink.org
  The Online Environmental Community
Environmental Defense Fund   www.edf.org
Environmental News Network   www.enn.com
Environmental Protection Agency   www.epa.gov
Environmental Working Group   www.ewg.org
Farm Service Agency   www.fsa.usda.gov
Fetzer Vineyards   www.fetzer.com
Flea Street Café   www.fleast.com
Food and Agriculture Organization   www.fao.org
  of the United Nations
Food and Drug Administration   www.fda.gov
Food First   www.foodfirst.org
Food Safety   www.foodsafety.org
Food Safety and Inspection Service   www.fss.usda.gov
Foreign Agricultural Service   www.fas.usda.gov
Green Gulch Farm Zen Center   http://bodhi.zendo.com
Health and Human Services   www.hhs.org
Health World Online   www.healthy.net
Heifer Project International   www.heifer.org
Henry A. Wallace Institute   www.hawiaa.org
for Alternative Agriculture
Institute for Agriculture and Trade Policy   www.iatp.org
International Federation   www.ecoweb.dk/ifoam
  of Organic Agriculture Movements
Intervale Foundation   www.gardeners.com
Kendall College Farm School   http://topaz.kenyon.edu
Land Stewardship Project   www.misa.umn.edu/lsphp
Leopold Center for Sustainable Agriculture   www.ag.iastate.edu/center/
                                             leopold
Mayo Clinic Health Oasis   www.maayohealth.org
Monsanto   www.monsanto.com
Mothers & Others for a Livable Planet   www.mothers.org
National Agriculture Library   www.nalusda.gov
National Gardening Association   www.garden.org
National Institutes for Health   www.nih.gov
National Marine Fisheries   www.nmfs.gov

National Organic Program	www.ams.usda.gov
National Resource Conservation Service	www.nrcs.usda.gov
National Resources Defense Council	www.nrdc.org
National Seed Storage Laboratory	www.ars-grin/nssl/nsslmain.html
Native Seeds/SEARCH	www.desert.net/seeds
NOFA of Vermont	www.nofavt.org
Northeast Cooperatives	www.northeastcoop.com
Old Chatham Shepherding Company Inn	www.oldsheepinn.com
Oldways Preservation and Exchange Trust	www.oldwayspt.org
Organic Farmers Marketing Association	www.iquest.net/ofma
Organic Farming Research Foundation	www.ofrf.org
Organic Trade Association	www.ota.com
Permaculture Site	http://nornet.nor.com.au/ enviornmental/perma.index.html
Pest Management at the Crossroads	www.pmac.net
Pesticide Action Network	www.panna.org
The Pesticide Trust	www.gn.apc.org
Public Citizen	www.citizen.org
Restaurant Nora	www.noras.com
RoxSand Restaurant and Bar	www.roxsand.com
Rural Advancement Foundation International	www.rafi.ca
Safe Tables Our Priority	www.stop-usa.org
Science News Online	www.sciencenews.org
SeaWeb	www.seaweb.org
Second Harvest	www.secondharvest.org
Share Our Strength	www.strength.org
Sustainable Agriculture Research and Education	www.sare.org
Sustainable Farming Connection	http://metalab.unc.edu/ farming~connection
Toxics & Waste	www.igc.org
Tyson Foods, Inc.	www.tyson.com
University of California Research and Education Program	www.sarep.ucdavis.edu
University of Missouri	www.ssu.missouri.edu
U.S. Census Bureau	www.census.gov
U.S. Water News Online	www.uswaternews.com
USDA—United States Department of Agriculture	www.usda.gov
USDA Agricultural Marketing Service	www.amsusda.gov
USDA Animal and Plant Inspection Services	www.aphis.usda.gov
Vermont Butter and Cheese Company	www.vtbutterandcheeseco.com

Vermont Department of Agriculture	www.state.vt.us/agric
Vermont Land Trust	www.vlt.org
Walnut Acres Organic Farms	www.walnutacres.com
Whole Foods	www.wholefoods.com
World Hunger Year	www.iglou.com/why
World Resources Institute	www.wri.org
World Sustainable Agriculture Association	www.igc.apc.org
Worldwatch Institute	www.worldwatch.org
Your Health	www.yourhealth.com

Notes

CHAPTER 1

1. Reay Tannahill, *Food in History* (New York: Crown, 1973), xv.
2. Dave Miller, quoted in "Women, Food and Technology," a lecture to members of the Culinary Historians of Boston, 9 April 1990. Cited by Barbara Wheaton in *Culinary Historians of Boston Newsletter*, fall 1989–1990.
3. Beverly Cox and Martin Jacobs, *Spirit of the Harvest: North American Indian Cooking* (New York: Stewart, Tabori & Chang, 1991), 142.
4. William Harlan Hale, *The Horizon Cookbook and Illustrated History of Eating and Drinking through the Ages* (New York: American Heritage, 1968), 14.
5. Tannahill, *Food in History*, 199–201.
6. Ibid., 218–19.
7. Ibid., 220.
8. Hiram M. Drache, *History of U.S. Agriculture and Its Relevance Today* (Danville, Ill.: Interstate Publishers, 1996), 25–30.
9. Ibid., v–vi.
10. James Trager, *The Food Chronology* (New York: Henry Holt, 1995), 172.
11. Ibid., 221.
12. Ibid., 232.
13. Drache, *History of U.S. Agriculture*, 133–35.
14. Trager, *Food Chronology*, 337.
15. Ibid., 381–82.
16. Ibid., 198.
17. Ibid., 460.
18. Ibid., 462.
19. Ibid., 480–85.
20. Drache, *History of U.S. Agriculture*, 269–73.
21. Mothers & Others for a Livable Planet, *The Green Food Shopper: An Activist's Guide to Changing the Food System* (New York: Mothers & Others for a Livable Planet, 1997), 32–33.
22. Trager, *Food Chronology*, 579.
23. Ibid., 587.

CHAPTER 2

1. *Progressive Grocer* (Darien Conn.), http://www.progressivegrocer.com/news.
2. Vermont Chapter of Chef's Collaborative 2000, letter to Vermont Chapter Member and the U.S. Department of Agriculture, April 1998.
3. Reed Karaim, "What's in a Label: Organic Hash from the USDA Kitchen," *Washington Post*, 30 March 1998, 23.
4. Patrick Leahy, interview by author, Washington, D.C., 23 September 1998.
5. Frederick Kirschenmann, "'Organic' May Not Be 'Sustainable,'" *Organic Farmer: The Digest of Sustainable Agriculture, Rural Education Action Project* 4, no. 1 (winter 1993): 18.
6. Joan Gussow, "Keynote Address: Getting There from Here" (presentation to Chefs Collaborative 2000 annual retreat, Puerto Rico, 21–24 January 1996).
7. Wes Jackson, *New Roots for Agriculture* (Lincoln: University of Nebraska Press, 1980), 14–21.
8. Michael Ableman, interview by author, Goleta, Calif., 27 July 1998.
9. Jackson, *New Roots for Agriculture*, 28.
10. The U.S. Salinity Laboratory, *www.ussl.ars.usda.gov*.
11. Hilary Baum, interview by author, New York, 15 July 1998.
12. Annie Berthold-Bond, *The Green Kitchen Handbook* (New York: HarperCollins, 1997), 32–33.
13. Martin Teitel, *Rain Forest in Your Kitchen* (Washington, D.C.: Island Press, 1992), 4.
14. National Seed Storage Laboratory, "Seed Viability and Storage Research Unit: Mission Statement," *National Seed Storage Laboratory*, 3 September 1997, http://checkers.nssl.colostate.edu/preserve/preserve.htm.
15. Kent Whealy, interview by author, Dubuque, Iowa, 21 July 1998.
16. William Woys Weaver, *Heirloom Vegetable Gardening: A Master Gardener's Guide to Planting, Seed Saving, and Cultural History* (New York: Henry Holt and Company, 1997), xxii.
17. Anne Raver, "A Vermont Hospital Turns its Garbage into Gold," *New York Times*, 9 October 1994.
18. Annie Somerville, interview by author, San Francisco, Calif., 30 July 1998.
19. Andrea Asch and Todd Kane, interview by author, South Burlington, Vt., 16 July 1998.
20. Teitel, *Rain Forest in Your Kitchen*, 63.
21. George Southworth, interview by author, Putney, Vt., 22 July 1998.
22. Michael Ableman, *On Good Land: The Autobiography of an Urban Farm* (San Francisco: Chronicle Books, 1998), 137.
23. Bob Anderson, interview by author, Penns Creek, Penn., 15 July 1998.
24. Jack L. Runyon, "A Profile of Hired Farm Workers, 1994 Annual Averages," USDA Economics Research Service, ERS-AER-748 (February 1997) http://www.econ.ag.gov.
25. Patricia Allen, *Food for the Future* (New York: John Wiley & Sons, 1993), 146.
26. Rick Bayless, interview by author, Chicago, Ill., 9 July 1998.
27. Eve Felder, interview by author, Hyde Park, N.Y., 20 July 1998.

28. Anderson, interview.

29. Cindy and David Major, interview by author, Putney, Vt., 6 August 1998.

30. Allison Hooper, interview by author, Websterville, Vt., 15 July 1998.

31. Bunny and Peter Flint, interview by author, Turnbridge, Vt., 21 July 1998.

32. Melvin D. Saunders, *Creative Alternatives for a Changing World* (Odenton, Md.: Creative Alternatives, 1997); www.braincourse.com/biodyna. html

33. Eliot Coleman, interview by author, Brooksville, Me., 23 July 1998.

34. Enid Wannacott, interview by author, Vt., 11 August 1998.

35. *Ben & Jerry's 1996 Ceres Report.*

36. *Ben & Jerry's 1997 Annual Report.*

37. Andrea Asch and Todd Kane, interview by author, South Burlington, Vt., 16 July 1998.

38. Ibid.

39. Southworth, interview.

40. Ableman, *On Good Land*, 7.

41. Ken Kelley, "Visions: Alice Waters," *Mother Jones*, January/February 1995, http://bsd.mojones.com/mother_jones/JF95/kelley.html.

42. Craig Wilson, "Chef Waters Plants Seeds of Environmentalism," *USA Today*, 3 June 1997, 12D.

43. Melissa Kelly, interview by author, Chatham, N.Y., 3 August 1998.

44. Tammy Lax, interview by author, Madison, Wis., 7–10 August 1998.

45. Odessa Piper, interview by author, Madison, Wis., 7–10 August 1998.

46. *Wine Spectator*, 30 April 1998

47. Michael Romano, interview by author, New York, 6 August 1998.

48. Ana Sortun, interview by author, Cambridge, Mass., 21 July 1998.

49. Deborah Madison and Edward Brown, *The Greens Cookbook* (New York: Bantam Books, 1987).

50. Somerville, interview.

51. Roger Clapp, interview by author, Montpelier, Vt., 7 July 1998.

52. Ibid.

53. Sara Baer-Sinnott, interview by author, Cambridge, Mass., 8 July 1998.

54. Tim Ryan, interview by author, Hyde Park, N.Y., 14 August 1998.

55. Felder, interview.

56. Jamie Eisenberg, interview by author, Montpelier, Vt., 10 July 1998.

57. K. Dun Gifford, interview by author, Cambridge, Mass., 8 July 1998.

58. Oldways website, http://www.oldwayspt.org/html/meet.htm.

59. Alice Waters, interview by author, Berkeley, Calif., 23 July 1998.

60. Mary Jo Viederman, interview by author, Londonderry, N.H., 3 August 1998.

61. Mothers & Others for a Livable Planet, "Mothers & Others' Shoppers' Campaign for Healthy Food, Farms, and Families," *Mothers & Others for a Livable Planet*, 15 December 1998, http://www.mothers.org/fieldwork/index.html.

62. Mothers & Others for a Livable Planet, "Core Values Northeast: Consumers. What is a Core Values Northeast Apple?" *Mothers & Others for a Livable Planet*, 12 December 1999, http://www.corevalues.org/consumers/ecolabel.html.

63. Catherine Sneed, interview by author, San Francisco, Calif., December 1998.
64. Darby Bradley, interview by author, Montpelier, Vt., 17 July 1998.
65. David Batcheldor, interview by author, Stratham, N.H., 6 August 1998.
66. Lester Brown et al., *State of the World 1998* (New York: W. W. Norton, 1998), 14–15.
67. Irena Chalmers, "A Search for a Sustainable Cuisine: Caribbean Conference On Sustainable Tourism" (presentation sponsored by the United Nations, November 29–December 2, 1995).
68. Dennis T. Avery and Alex Avery, "Farming to Sustain the Environment," Hudson Briefing Paper, *Shaping the Future*, no. 190 (May 1996): 1–20.
69. Erik P. Eckholm, *Losing Ground* (New York: W. W. Norton, 1976), 179–81.
70. Leahy, interview.
71. Second Harvest, *Hunger 1997: The Faces and Facts* (Chicago: Second Harvest, 1997), 8.
72. Jenny Tesar, *Food and Water: Threats and Shortages and Solutions* (New York: Facts on File, 1992), 9–10.

CHAPTER 3

1. Joan Dye Gussow, "What Is Sustainable Cuisine—and Why Does It Matter So Much?" (presentation to 5th Annual Chef's Collaborative 2000 retreat, Phoenix, Ariz., 7–9 June 1998).
2. Martin Teitel, *Rain Forest in Your Kitchen* (Washington, D.C.: Island Press, 1992), 32.
3. Elizabeth Henderson, "Thoughts Still Relevant for a Summer Day: Keynote from the NOFA Vermont Winter Conference," *The Beet* (Putney Co-op, Putney, Vt.), July/August 1998, 2.
4. David C. Korten, *When Corporations Rule the World* (San Francisco: Kumarian Press and Berrett-Koehler Publishers, 1995), 224–25.
5. Scott Kilman and Susan Warren, "Old Rivals Fight for New Turf—Biotech Crops," *Wall Street Journal*, 27 May 1998, B1, B7.
6. Ibid.
7. Teitel, *Rain Forest in Your Kitchen*, x.
8. Monsanto Company press release, 29 September 1998, www.monsanto.com.
9. Acres U.S.A., " 'Terminator Technology' Threatens Future World Food Supply," *Acres U.S.A.—A Voice for Eco-Agriculture* 28, no. 12 (December 1998): 17.
10. Joan Gussow, interview by author, Piermont, N.Y., 7 July 1998.
11. Acres U.S.A., " 'Terminator Technology.' "
12. Ibid.
13. The Bruntland Report to which Phil refers was given in 1987 by the World Commission on Environment and Development. It defines sustainable development as progress that "meets the needs of the present without compromising the ability of the future generations to meet their own needs."
14. Phil Angel, interview by author, St. Louis, Mo., 27 October 1998.
15. Patrick Leahy, interview by author, Washington, D.C., 23 September 1998.

16. Michael Greger, "Bovine Growth Hormone—What You May Not Know," *Animal Rights Resource Link* (Retrieved 28 October 1998), http://arrs. envirolink.org/AnimalLife/spring95/BGH.html.
17. www.monsanto.com/monsanto/financial/annual.
18. Mental Health Infosource, "Statistics Related to Overweight and Obesity," Continuing Medical Education Inc., July 1996 (retrieved January 1998), www.mhsource.com/hy/obstats.html.
19. Monsanto Company, *1997 Annual Report* (St. Louis: Monsanto Company, 1997), 2.
20. Ibid., 8–12.
21. Marc Lappé and Britt Bailey, *Against the Grain: Biotechnology and the Corporate Takeover of Food* (Monroe, Me.: Common Courage Press, 1998), 5.
22. Teitel, *Rain Forest in Your Kitchen*, 59–62.
23. Ibid., 41.
24. "Rodale 15-Year Trials: Organic Farming Finishes First," *Wholistic News Magazine*, May/June 1997, 6.
25. Jenny Tesar, *Food and Water: Threats and Shortages and Solutions* (New York: Facts on File, 1992), 43.
26. Michael Ableman, *From the Good Earth: A Celebration of Growing Food around the World* (New York: Harry N. Abrams, 1993), 73.
27. Stephen M. Voynick, *Riding the Higher Range: The Story of Colorado's Coleman Ranch and Coleman Natural Beef* (Saguache, Colo.: Glenn Melvin Coleman, 1998), 11.
28. Ibid., 242.
29. Jeremy Rifkin, *The Biotech Century: Harnessing the Gene and Remaking the World* (New York: Putnam, 1998), 17–20.
30. Gary Paul Nabhan, *Enduring Seeds: Native American Agriculture and Wild Plant Conservation* (New York: North Point Press, 1989), 159–62.
31. Ibid.
32. Ibid., 162.
33. Richard Roop, interview by author, Springdale, Ariz., 27 October 1998.
34. Frederick Golden, "Getting a Leg Up on the Birds," *Time*, 10 May 1999, 73.
35. Stan Grossfeld, "Animal Waste Emerging as US Problem, Part 2 of 3," *Boston Globe*, 21 September 1998, A1, 8–9.
36. Betsy Freese, "Successful Farming Exclusive: Pork Powerhouses, 1997 Largest Farms Keep Expanding but Disease Hurts Sow Productivity," Meredith Corporation, Des Moines, website, October 1997 (retrieved 12 December 1999), http://www.agriculture.com/sfonline/sf/1997/october/pork97/index.html.
37. Mark Kurlansky, "Catch of the Day: Stocks of Many of Our Favorite Fish Are in Trouble. Why Don't We Care?" *Food & Wine*, December 1997, 90.
38. Steve Connolly, interview by author, Boston, Mass., 31 August 1998.
39. Carl Safina, *Song for the Blue Ocean Encounters along the World's Coasts and Beneath the Seas* (New York: Henry Holt, 1997), 30.
40. "Give Swordfish a Break," SeaWeb and Natural Resources Defense Council website, January 1998, http://www.seaweb.org/swordfish.

41. Doug Raider, interview by author, Raleigh, N.C., 21 July 1998.
42. Mark Kurlansky, *Cod: A Biography of the Fish That Changed the World* (New York: Walker and Company, 1997), 232.
43. "North Atlantic Salmon: End of the Line," *Economist*, 14 February 1998, 80.
44. Lester R. Brown et al., *State of the World 1998* (New York: W. W. Norton, 1998).
45. David Wills, interview by author, Washington, D.C., 14 July 1998.
46. Claude E. Boyd and Jason W. Clay, "Shrimp Aquaculture and the Environment: An Adviser to Shrimp Producers and an Environmentalist Present a Prescription for Raising Shrimp Responsibly," *Scientific American*, June 1998, 58–65.
47. Jim Carlberg, interview by author, San Diego, Calif., 3 August 1998.
48. Romano, interview.
49. Rifkin, *The Biotech Century*, 22.
50. Angel, interview.
51. Rifkin, *The Biotech Century*, 50–51.
52. Angel, interview.
53. Rifkin, *The Biotech Century*, 80–86.
54. Brown et al., *State of the World 1998*, 53.
55. Angel, interview.
56. Lappé and Bailey, *Against the Grain*, 50.
57. Rifkin, *The Biotech Century*, 231–33.
58. Kevin Knox, interview by author, Putney, Vt., 13 July 1998.
59. National Fisheries Institute, website www.nfi.org/industr5.html.
60. Brown et al., *State of the World 1998*, 68–71.
61. Korten, *When Corporations Rule the World*, 30.
62. Lappé and Bailey, *Against the Grain*, 76–77.
63. Rifkin, *The Biotech Century*, 103–5.
64. "Monsanto Requests Delay on GE Crops In USDA Organic Standards," Organic Farmers Marketing Association (OFMA) website, 17 April 1998, http://web.iquest.net/ofma/monsanto.htm.
65. Angel, interview.
66. Lappé and Bailey, *Against the Grain*, 132.
67. Ibid., 91.

CHAPTER 4

1. Daniel P. Puzo, "Crisis Prevention?" *Restaurants and Institutions*, 15 January 1998.
2. Al Gore, *Earth in the Balance: Ecology and the Human Spirit* (New York: Plume, 1993), 1–6.
3. PRNewswire, Washington, D.C., 16 March 1999, www.prnewswire.com.
4. Nichols Fox, *Spoiled: Why Our Food Is Making Us Sick and What We Can Do About It* (New York: Penguin, 1997), 47.
5. Stan Grossfeld, "New Dangers Make Way to US Tables," *Boston Globe*, 20 September 1998, A1, 30, 31.

6. John Wargo, *Our Children's Toxic Legacy: How Science and Law Fail to Protect Us From Pesticides* (New Haven: Yale University Press, 1998), 162–63.

7. Gerry Clark, "Clinton Creates Council on Food Safety," *Food Product Design*, October 1998, 16.

8. Karen Springen, "Safer Food for a Tastier Millennium," *Newsweek*, 28 September 1998, 14.

9. Business Wire, Athens, Ga., 27 January 1999.

10. "Food Safety Concerns Spur New Coalitions, Increased Education, and Professional Certification," *National Culinary Review*, January 1998, 10–11.

11. Donna U. Vogt, "Food Additive Regulations: A Chronology," *Congressional Research Service: Report for Congress*, 95–857 SPR, Washington, D.C., 13 September 1995.

12. *Green Guide* (New York), no. 67 (May 1999).

13. Jenny Tesar, *Food and Water: Threats and Shortages and Solutions* (New York: Facts on File, 1992), 47.

14. Ibid., 50.

15. Wargo, *Our Children's Toxic Legacy*, 94.

16. Ibid., 101.

17. Ibid., 301.

18. Ibid., 302.

19. Ibid., 7.

20. Michael Ableman, *From the Good Earth: A Celebration of Growing Food around the World* (New York: Harry N. Abrams, 1993), 74–76.

21. Wargo, *Our Children's Toxic Legacy*, 101.

22. Robin Lee Allen, "Irradiation Approved by FDA as Safety Tool: Public's Acceptance a Question," *Nation's Restaurant News*, 15 December 1997, 1.

23. Fred Genth, interview by author, Greeley, Colo., 11 November 1998.

24. Fox, *Spoiled*, 138.

25. Ibid., 167–68.

26. Ibid., 178–79.

27. Ibid., 252.

28. "Study: Cattle Diet Change Could Cut E. Coli Risk," Associated Press, 11 September 1998.

29. David C. Korten, *When Corporations Rule the World* (San Francisco: Kumarian Press and Berrett-Koehler, 1995), 146.

30. Janelle Carter, "Diet Guidelines May Add Food Safety," Associated Press, 9 March 1999.

31. Peter Annin, "Defeat at Pork Chop Hill," *Newsweek*, 21 December 1998, 48.

32. Donald L. Barlett and James B. Steele, "The Empire of the Pigs," *Time*, 30 November 1998.

33. U.S. Food and Drug Administration Center for Food Safety and Applied Nutrition Office of Seafood website, http://vm.cfsan.fda.gov/~dms/haccp.

34. "HACCP Compliance Is Low, but Most Infractions Are Minor," *Seafood Business*, February 1999, 1.

35. Patricia Allen, *Food for the Future* (New York: John Wiley & Sons, 1993), 142–43.
36. *How SARE Works: SARE 1998 Project Highlights* (Washington, D.C.: 1998), Sustainable Agriculture Research and Education, 2.
37. Miranda Smith and Elizabeth Henderson, eds., *The Real Dirt: Farmers Tell About Organic and Low-Input Practices in the Northeast* (Burlington, Vt.: Northeast Region Sustainable Agriculture Research and Education Program, 1998).
38. Anne Witte Garland with Mothers & Others for a Livable Planet, *The Way We Grow: Good-Sense Solutions for Protecting Our Families from Pesticides in Food* (New York: Berkeley Books, 1993), 35.
39. Paul Hawken, *The Ecology of Commerce* (New York: Harper Business, 1993), 114.
40. Al Gore, *Earth in the Balance: Ecology and the Human Spirit* (New York: Plume, 1993), 167.
41. Farm Service Agency Online website, www.fsa.usda.gov/daco/default.asp.
42. "Clinton Announces Food Aid Plan Aimed at Boosting US Farmers," Agence France-Presse website, 18 July 1998 (retrieved 19 July 1998), http://library.northernlight.com/GG19980727050001056.html?cb=0&sc=0#doc.
43. Lester R. Brown et al., *State of the World 1998* (New York: W. W. Norton, 1998), 91.
44. Ibid., 91, 94–95.
45. Korten, *When Corporations Rule the World*, 179.
46. Ibid.
47. Ibid.
48. Suzie Larsen, "Update: Pesticide Dumping Continues; Leahy to Reintroduce Circle of Poison Bill in the Clinton Era, U.S. Dumping of Hazardous Pesticides Overseas Has Gotten Worse," *Mother Jones Magazine*, 21 July 1998, http://bsd.mojones.com/news_wire/pest_dump.html.
49. Mothers & Others for a Livable Planet, *The Green Food Shopper: An Activist's Guide to Changing the Food System* (New York: Mothers & Others for a Livable Planet, 1997), 87–88.
50. Larsen, "Update."
51. Ableman, *From the Good Earth*, 77–79.
52. www.citizen.org/pctrade/nafta/reports/trick.
53. Wargo, *Our Children's Toxic Legacy*, 162–63.
54. Public Citizen Global Trade Watch NAFTA website, www.citizen.org/pctrade/nafta/reports/trick.
55. "President Clinton Announces Initiative to Ensure the Safety of Imported and Domestic Fruits and Vegetables," Office of the Government White House website, 2 October 1997, http://vm.cfsan.fda.gov/~dms/fsfact2.html.
56. USDA Food, Nutrition and Consumer Services website, www.fns.usda.gov/fncs.
57. USDA FNS online, Food Stamps website www.fns.usda.gov/fsp.

58. USDA FNS online, Women, Infants and Children website, www.fns.usda. gov/wic.

59. USDA FNS online, School Breakfast Program website, www.fns.usda.gov/ cnd/breakfast.

60. USDA FNS online, Summer Food Service Program website, www.fns.usda. gov/cnd/summer.

61. Ellen Haas, interview by author, Washington, D.C., 14 July 1998.

CHAPTER 5

1. Paul Recer, "Americans Heading toward Fat Trend," Associated Press, 28 March 1998.

2. Martin Teitel, *Rain Forest in Your Kitchen* (Washington, D.C.: Island Press, 1992), 90–92.

3. Laura Shapiro, "In Sugar We Trust," *Newsweek*, 13 July 1998, 72–74.

4. "Statistics Related to Overweight and Obesity," Mental Health Infosource website, Continuing Medical Education Inc., Irvine, Calif., July 1996 (retrieved January 1998), www.mhsource.com/hy/obstats.html.

5. Malcolm Gladwell, "Excerpt from 'Annals of Medicine: The Pima Paradox,'" *New Yorker Electronic Newsstand*, 2 February 1998, http://magazines.enews.com/magazines/new_yorker/archive/980202–001.html.

6. Terry Shintani, excerpt from *The Hawaiian Paradox* (1993), in *Food Choices: 2000 Sustainable Diets for the Next Century: An Oldways International Symposium Schedule of Sessions & Events, Background Papers & Information Hawaii—10–15 July* (Cambridge, Mass.: Oldways, 1993), 53.

7. Alexandra Marks, "America's Changing Dinner Plate," *Christian Science Monitor*, 8 August 1997, http://video.csmonitor.com/durable/1997/08/08/us/us.5.html.

8. Laura Pensiero et al., *The Strang Cookbook for Cancer Prevention* (New York: Dutton, 1998), xv.

9. Laura Pensiero, interview by author, Hyde Park, N.Y., 18 December 1998.

10. Geoffrey Cowley, "Cancer and Diet," *Newsweek*, 30 November 1998.

11. "Tomatoes May Help Ward Off Cancer, Institute Says," *Nation's Restaurant News* 33, no. 12 (22 March 1999): 44.

12. Ibid.

13. Jack Kittredge, "News Notes," *The Natural Farmer* (Northeast Organic Farming Association, Barre, Mass. 2) no. 40 (Spring 1999): 1.

14. Rebecca Goldberg and Tracy Triplett, *Murky Waters: Environmental Effects of Aquaculture in the United States* (New York: Environmental Defense Fund, 1997), 31.

15. "Organic Foods: Safer? Tastier? More Nutritious?" *Consumers Union Magazine*, Consumer Reports Online website, January 1998, http://www.consumerreports.org/Categories/FoodHealth/Reports/9801or g1.htm.

16. Guy Dauncey, *EcoNews* (Victoria, British Columbia, Canada), retrieved 27 June 1998, http://www.islandnet.com/~gdauncey/econews.

17. Patricia Allen, *Food for the Future* (New York: John Wiley & Sons, 1993), 48.

18. Ibid.

19. Deborah Branscum, "Rebuilding the Pyramid: Innovative Educators Help Kids Tackle Bad Eating Habits," *Eating Well*, March 1998, 88.

20. "Diet Therapy for Attention Deficit Hyperactivity Disorder Gains Support By Doctors, Researchers And Parents," PRNewswire, Riverhead, N.Y., 10 April 1996.

21. John Wargo, *Our Children's Toxic Legacy: How Science and Law Fail to Protect Us from Pesticides* (New Haven: Yale University Press, 1998), 5.

22. Sandra Steingraber, *Living Downstream* (Reading, Mass.: Perseus Books, 1997), 164–68.

23. Wargo, *Our Children's Toxic Legacy*, 104.

24. Steingraber, *Living Downstream*, 164–68.

25. Dan Goldberg, "The Godzilla Effect: Way Bigger, Not Way Better," *Curmudgeon's Home Companion: Dan Goldberg's Monthly Guide to Reality and Good Food*, June 1998, 1.

26. "Home Meal Replacement Study Reveals Consumer Eating Habits USA Chicago, Inc., Surveys 25,000 Area Households," PRNewswire, 23 March 1998.

27. Irena Chalmers, "A Search for a Sustainable Cuisine: Caribbean Conference On Sustainable Tourism" (presentation sponsored by the United Nations, 29 November 29 to 2 December 1995).

28. David C. Korten, *When Corporations Rule the World* (San Francisco: Kumarian Press and Berrett-Koehler, 1995), 154–55.

29. "In the News: A Snapshot of Food and Nutrition Reporting," *Food Insight Newsletter* (IFIC Foundation, National Food Service Management Institute [NFSMI], University of Mississippi), January/February 1998, 1, 4–5.

30. Jean Seligmann with Sarah Van Boven, "Got Soy Milk?" *Newsweek*, 29 June 1998, 67.

31. Laura Shapiro, "Those Chips Look Good. But Do You Lose Weight?" *Newsweek*, 29 June 1998, 68.

32. Greg Johnson, "Olestra Builds Fat Niche in Snack Market; Products with the New, No-Calorie Additive Are Selling Strongly, Aided by FDA Blessings," *Los Angeles Times*, 17 July 1998. http://www.latimes.com/HOME/ARCHIVES/simple.htm?bot.

33. "FDA OKs New Artificial Sweetener," Associated Press, 2 April 1998.

34. Michael Ableman, *On Good Land: The Autobiography of an Urban Farm* (San Francisco: Chronicle Books, 1998), 75.

35. Lynn A. Kuntz, "Functional Food Fight," *Food Product Design*, February 1999, 12.

36. Christine Gorman, "It Sure Ain't Butter," Online Time Health Column website, May 28, 1999.

37. Goldberg and Triplett, *Murky Waters*, 43.

38. Ibid., 45.

39. Dick Thompson, "Drugged Chicks Hatch a Menace," *Time*, 31 May 1999.

40. Denise Grady, "FDA Revising Guidelines for Antibiotics for Animals," *New York Times*, 8 March 1999.

41. Ellen Ruppell Shell, "Could Mad Cow Disease Happen Here?" *Atlantic Monthly*, September 1998, 92–94, 97,98, 100–2, 104–6.

42. Jim Avila, "Dire Warning on Lunch Meat, Chicago," *NBC News*, 14 April 1999, www.msnbc.com.

43. Herman-Giddens ME, Slora E. J., Wasserman R. C. et al. "Secondary Sexual Characteristics and Menses in Young Girls Seen in Office Practice: A Study from the Pediatric Research in Office Settings Network," *Pediatrics* (1997): 505–12.

44. Kristin Ebbert and Becky Gillette, "Hormonal Imbalance: How Pollution Skews Sexual Development," *The Green Guide* (Mothers & Others for a Livable Planet, New York), 1 June 1998, 1, 4.

45. Mark Alpert, "Where Have All the Boys Gone? The Mysterious Decline in Male Births," *Scientific American*, July 1998, www.sciam.com/1998/0798issue/0798scicit3.html.

46. Rachel Carson, *Silent Spring* (New York: Houghton Mifflin, 1962), xxii–xxiii.

47. Christopher Drew and Pam Belluck, "Deadly Bacteria a New Threat to Fruit and Produce in U.S.," *New York Times*, 4 January 1998.

48. Susan G. Strother, "Restaurant Group Blames Most Food Poisonings on Victims' Own Kitchens," *Orlando Sentinel*, 29 August 1998, http://www.orlandosentinel.com/archives.

49. Nichols Fox, *Spoiled: Why Our Food Is Making Us Sick and What We Can Do about It* (New York: Penguin, 1997), 91–92.

50. Ibid., 358–59.

51. Ibid., 72–74.

52. David Mas Masumoto, *Epitaph for a Peach: Four Seasons on My Family Farm* (San Francisco: HarperCollins, HarperSanFrancisco: 1995), ix.

CHAPTER 6

1. Paul Hawken, *The Ecology of Commerce* (New York: Harper Business, 1993), 159–60.

2. Eliot Coleman, *The New Organic Grower*, 2d. ed. (White River Junction, Vt.: Chelsea Green Publishing, 1995), 293.

3. Rebecca Goldberg and Tracy Triplett, *Murky Waters: Environmental Effects of Aquaculture in the United States* (New York: Environmental Defense Fund, 1997), 13.

4. Dick Monroe, interview by author, Orlando, Fla., 3 August 1998.

5. Jim Carlberg, interview by author, San Diego, Calif., 3 August 1998.

6. Leonard Engle, excerpt from *The Sea* (1961), "Ocean Food Chain: Life From Death," in *Food Choices: 2000 Sustainable Diets for the Next Century: An Oldways International Symposium Schedule of Sessions & Events, Background Papers & Information—Hawaii July 10–15* (Cambridge, Mass.: Oldways, 1993), 24.

7. David Wills, interview by author, Washington, D.C., 14 July 1998.

8. Laura Fraser, "Homegrown Harvest," *Health*, 6 May 1997, 70.

9. Irena Chalmers, "A Search for a Sustainable Cuisine: Caribbean Conference on Sustainable Tourism" (presentation sponsored by the United Nations, 29 November to 2 December 1995).

10. Miranda Smith and Elizabeth Henderson, eds., *The Real Dirt: Farmers Tell About Organic and Low-Input Practices in the Northeast* (Burlington,

Vt.: Northeast Region Sustainable Agriculture Research and Education Program, University of Vermont, 1998), 185.

11. Rick Bayless, interview by author, Chicago, Ill., 9 July 1998.
12. Odessa Piper, interview by author, Madison, Wis., 7–10 August 1998.
13. Vern Grubinger, interview by author, Brattleboro, Vt., 21 December 1998.
14. Eve Felder, interview by author, Hyde Park, N.Y., 20 July 1998.
15. Hawken, *Ecology of Commerce*, 184.
16. Joyce Goldstein, interview by author, San Francisco, Calif., 29 July 1998.
17. Bayless, interview.

Recommended Reading

Ableman, Michael. *From the Good Earth: A Celebration of Growing Food around the World.* New York: Harry N. Abrams, 1993.

————. *On Good Land: The Autobiography of an Urban Farm.* San Francisco: Chronicle Books, 1998.

Allen, Patricia. *Food for the Future.* New York: John Wiley & Sons, 1993.

Arnold, Andrea. *Fear of Food: Environmentalist Scams, Media Mendacity, and the Law of Disparagement.* Bellevue, Wash.: Free Enterprise Press, 1990.

Ash, John, and Sid Goldstein. *From the Earth to the Table—John Ash's Wine Country Cuisine.* New York: Dutton/Signet, 1996.

Berry, Wendell. *The Unsettling of America: Culture and Agriculture.* San Francisco: Sierra Club Books, 1986.

Berthold-Bond, Annie. *The Green Kitchen Handbook.* New York: Harper-Collins, HarperPerennial, 1997.

Better Homes and Garden Books. *Better Homes & Gardens Heritage Cookbook.* Meredith, 1975.

Brown, Lester R., and Hal Kane. *Full House: Reassessing the Earth's Population Carrying Capacity.* New York: W. W. Norton, 1994.

Brown, Lester R., et al. *State of the World 1998.* New York: W. W. Norton, 1998.

————. *Vital Signs 1998: The Environmental Trends That Are Shaping Our Future.* New York: W. W. Norton, 1998.

Cairncross, Frances. *Costing the Earth: The Challenge for Governments, the Opportunities for Business.* Boston: Harvard Business School Press, 1993.

Carson, Rachel. *Silent Spring.* New York: Houghton Mifflin, 1962.

Coleman, Eliot. *Four-Season Harvest.* White River Junction, Vt.: Chelsea Green Publishing, 1992.

————. *The New Organic Grower.* 2d. ed. White River Junction, Vt.: Chelsea Green Publishing, 1995.

Cool, Jesse Ziff. *Onions: A Country Garden Cookbook.* New York: Harper-Collins, 1995.

————. *Tomatoes: A Country Garden Cookbook.* New York: HarperCollins, 1994.

Degler, Teri, and Pollution Probe. *The Kitchen Handbook: An Environmental Guide.* Toronto: McClelland & Stewart, 1992.

Drache, Hiram M. *History of U.S. Agriculture and Its Relevance Today.* Danville, Ill.: Interstate Publishers, 1996.

Eckholm, Erik P. *Losing Ground.* New York: W. W. Norton, 1976.

Eide, A., et al., eds. *Food as a Human Right.* Tokyo: United Nations University, 1984.

Fox, Nichols. *Spoiled: Why Our Food Is Making Us Sick and What We Can Do About It.* New York: Penguin, 1997.

Garland, Anne Witte, with Mothers & Others for a Livable Planet. *The Way We Grow: Good-Sense Solutions for Protecting Our Families from Pesticides in Food.* New York: Berkeley Books, 1993.

Goldberg, Rebecca, et al. *Biotechnology's Bitter Harvest: Report of the Biotechnology Working Group.* Washington, D.C.: Biotechnology Working Group, 1990.

Goldberg, Rebecca, and Tracy Triplett. *Murky Waters: Environmental Effects of Aquaculture in the United States.* New York: Environmental Defense Fund, 1997.

Goldstein, Joyce. *Kitchen Conversations.* New York: William Morrow, 1977.

———. *The Mediterranean Kitchen.* New York: William Morrow, 1989.

Gore, Al. *Earth in the Balance: Ecology and the Human Spirit.* New York: Plume, 1993.

Groh, Trauger, and Steven McFadden. *Farms of Tomorrow Revisited: Community Supported Farms, Farm Supported Communities.* Kimberton, Penn.: Biodynamic Farming and Gardening Association, 1997.

Gussow, Joan Dye. *Chicken Little, Tomato Sauce and Agriculture: Who Will Produce Tomorrow's Food?* New York: The Bootstrap Press, 1991.

Haas, Ellen. *Great Adventures in Food: Fresh Ways to Celebrate Every Meal.* New York: Golden Books, 1999.

Hale, William Harlan. *The Horizon Cookbook and Illustrated History of Eating and Drinking through the Ages.* New York: American Heritage, 1968.

Hawken, Paul. *The Ecology of Commerce.* New York: Harper Business, 1993.

Hirt, Paul W. *A Conspiracy of Optimism: Management of the National Forests since World War Two.* Lincoln: University of Nebraska Press, 1994.

Jackson, Wes. *New Roots for Agriculture.* Lincoln: University of Nebraska Press, 1980.

Jeavons, John. *How to Grow More Vegetables on Less Land Than You Can Imagine.* Berkeley, Calif.: Ten Speed Press, 1995.

Junger, Sebastian. *The Perfect Storm: A True Story of Men against the Sea.* New York: W. W. Norton, 1997.

Kiefer, Joseph, and Martin Kemple. *Digging Deeper.* Montpelier, Vt.: Common Roots Press, 1998.

Knox, Kevin, and Julie Sheldon Huffaker. *Coffee Basics: A Quick and Easy Guide.* New York: John Wiley & Sons, 1997.

Korten, David C. *When Corporations Rule the World.* San Francisco: Kumarian Press and Berrett-Koehler, 1995.

Kurlansky, Mark. *Cod: A Biography of the Fish That Changed the World.* New York: Walker and Company, 1997.

Lappé, Frances Moore, and Paul Martin DuBois. *The Quickening of America: Rebuilding Our Nation, Remaking Our Lives.* San Francisco: Jossey-Bass, 1994.

Lappé, Marc, and Britt Bailey. *Against the Grain: Biotechnology and the Corporate Takeover of Food.* Monroe, Me.: Common Courage Press, 1998.

Masumoto, David Mas. *Epitaph for a Peach: Four Seasons on My Family Farm.* San Francisco: HarperCollins, HarperSanFrancisco, 1995.

Mothers & Others for a Livable Planet. *The Green Food Shopper: An Activist's Guide to Changing the Food System.* New York: Mothers & Others for a Livable Planet, 1997.

———. *Mothers & Others Guide to Natural Baby Care.* New York: John Wiley & Sons, 1999.

———. *The Green Kitchen Handbook.* San Francisco: HarperCollins, 1997.

———. *A Report on Green Food Labels: Emerging Opportunities for Environmental Awareness and Market Development.* New York: Mothers & Others for a Livable Planet, 1996.

Nabhan, Gary Paul. *Enduring Seeds: Native American Agriculture and Wild Plant Conservation.* New York: North Point Press, 1989.

Pensiero, Laura, R.D., et al. *The Strang Cookbook for Cancer Prevention.* New York: Dutton, 1998.

Pouillon, Nora. *Cooking with Nora.* New York: Random House, 1996.

Revel, Jean François. *Culture and Cuisine.* New York: Doubleday, 1982.

Rich, Bruce. *Mortgaging the Earth: The World Bank, Environmental Impoverishment, and the Crisis of Development.* Boston: Beacon Press, 1994.

Rifkin, Jeremy. *The Biotech Century: Harnessing the Gene and Remaking the World.* New York: Putnam, 1998.

Roelofs, Joan. *Greening Cities: Building Just and Sustainable Communities.* New York: Bootstrap Press, 1996.

Romano, Michael. *The Union Square Café Cookbook.* San Francisco: Harper-Collins, 1994.

Safina, Carl. *Song for the Blue Ocean Encounters along the World's Coasts and beneath the Seas.* New York: Henry Holt, 1997.

Saunders, Melvin D. *Creative Alternatives for a Changing World.* Odenton, Md.: Creative Alternatives, 1997.

Somerville, Annie. *Fields of Greens: New Vegetarian Recipes from the Celebrated Greens Restaurant.* New York: Bantam Doubleday, 1993.

Stauber, John, and Sheldon Rampton. *Toxic Sludge Is Good for You: Lies, Damn Lies, and the Public Relations Industry.* Monroe, Me.: Common Courage Press, 1995.

Steingraber. Sandra. *Living Downstream.* Reading, Mass.: Perseus Books, 1997.

Tannahill, Reay. *Food in History.* New York: Crown, 1973.

Teitel, Martin. *Rain Forest in Your Kitchen.* Washington, D.C.: Island Press, 1992.

Tesar, Jenny. *Food and Water: Threats and Shortages and Solutions.* New York: Facts on File, 1992.

Tokar, Brian. *Earth for Sale: Reclaiming Ecology in the Age of Corporate Greenwash.* Boston: South End Press, 1997.

Toussaint-Samat, Maguelonne. *History of Food*. Translated by Anthea Bell. Cambridge, Mass.: Blackwell, 1987.

Trager, James. *The Enriched, Fortified, Concentrated, Country-Fresh, Lip-Smacking, Finger-Licking, International, Unexpurgated Food Book*. New York: Grossman, 1970.

———. *The Food Chronology*. New York: Henry Holt, 1995.

Voynick, Stephen M. *Riding the Higher Range: The Story of Colorado's Coleman Ranch and Coleman Natural Beef*. Saguache, Colo.: Glenn Melvin Coleman, 1998.

Wargo, John. *Our Children's Toxic Legacy: How Science and Law Fail to Protect Us from Pesticides*. New Haven, Conn.: Yale University Press, 1998.

Waters, Alice. *The Chez Panisse Café Cookbook*. San Francisco: HarperCollins, 1999.

———. *Chez Panisse Menu Cookbook*. New York: Random House, 1982.

———. *Chez Panisse Vegetables*. San Francisco: HarperCollins, 1996.

Weaver, William Woys. *Heirloom Vegetable Gardening: A Master Gardener's Guide to Planting, Seed Saving, and Cultural History*. New York: Henry Holt, 1997.

Index